Novel 3D Media Technologies

Ahmet Kondoz • Tasos Dagiuklas
Editors

Novel 3D Media Technologies

 Springer

Editors
Ahmet Kondoz
Institute for Digital Technologies
Loughborough University in London
Queen Elizabeth Olympic Park, London
United Kingdom

Tasos Dagiuklas
School of Science and Technology
Hellenic Open University
Patras, Greece

ISBN 978-1-4939-5453-7   ISBN 978-1-4939-2026-6 (eBook)
DOI 10.1007/978-1-4939-2026-6
Springer New York Heidelberg Dordrecht London

Printed on acid-free paper

Springer is part of Springer Science+Business Media (www.springer.com)

Contents

Contributors

Harsha D. Appuhami Kingston University, London, UK

Marcus Barkowsky LUNAM Université, Université de Nantes, IRCCyN UMR CNRS 6597, Polytech Nantes, Nantes, France

Athina Bourdena Department of Informatics Engineering, Technological Educational Institute of Crete, Heraklion, Crete, Greece

Marco Cagnazzo Institut Mines-Télécom, Télécom ParisTech, CNRS LTCI, Paris, France

Patrick Le Callet LUNAM Université, Université de Nantes, IRCCyN UMR CNRS 6597, Polytech Nantes, Nantes, France

Caroline Conti ISCTE - Instituto Universitário de Lisboa/Instituto de Telecomunicações, Lisbon, Portugal

Tasos Dagiuklas School of Science and Technology, Hellenic Open University, Patras, Greece

Carl J. Debono Department of Communications and Computer Engineering, University of Malta, Msida, Malta

Antoine Dricot Orange Labs, Guyancourt, France

Institut Mines-Télécom; Télécom ParisTech; CNRS LTCI, Paris, France

Frederic Dufaux Institut Mines-Télécom; Télécom ParisTech; CNRS LTCI, Paris, France

Sérgio M.M. Faria ESTG, Instituto Politécnico de Leiria/Instituto de Telecomunicações, Leiria, Portugal

Claudio Feijóo CeDInt-UPM. Campus de Montegancedo, Pozuelo de Alarcón, Madrid, Spain

Iris Galloso CeDInt-UPM. Campus de Montegancedo, Pozuelo de Alarcón, Madrid, Spain

Chaminda T.E.R. Hewage Kingston University, London, UK

Joel Jung Orange Labs, Guyancourt, France

Panagiotis Karkazis Electrical & Computer Engineering Department, Technical University of Crete, Chania, Greece

Ahmet Kondoz Institute for Digital Technologies, Loughborough University in London, Queen Elizabeth Olympic Park, London, UK

Ioannis Koufoudakis Synelixis Solutions Ltd, Chalkida, Greece

Helen C. Leligou Electrical Engineering Department, Technical Institute of Sterea Ellada Psachna, Psachna, Greece

Jing Li LUNAM Université, Université de Nantes, IRCCyN UMR CNRS 6597, Polytech Nantes, Nantes, France

Hyun Lim Institute for Digital Technologies, Loughborough University in London, Queen Elizabeth Olympic Park, London, UK

Evariste Logota Instituto de Telecomunicações, Aveiro, Portugal

Hugo Marques Instituto Politécnico de Castelo Branco, Castelo Branco, Portugal

Maria G. Martini Kingston University, London, UK

George Mastorakis Department of Informatics Engineering, Technological Educational Institute of Crete, Heraklion, Crete, Greece

Constandinos X. Mavromoustakis Computer Science Department, University of Nicosia, Nicosia, Cyprus

Paraskevi Mousicou Computer Science Department, University of Nicosia, Nicosia, Cyprus

Paulo Nunes ISCTE - Instituto Universitário de Lisboa/Instituto de Telecomunicacões, Lisbon, Portugal

Evangelos Pallis Department of Informatics Engineering, Technological Educational Institute of Crete, Heraklion, Crete, Greece

Katerina Papanikolaou European University Cyprus, Nicosia, Cyprus

Béatrice Pesquet Institut Mines-Télécom; Télécom ParisTech; CNRS LTCI, Paris, France

Christos Politis Kingston University, London, UK

Haroon Qureshi Institut für Rundfunktechnik GmbH (IRT), Munich, Germany

Nuno M.M. Rodrigues ESTG, Instituto Politécnico de Leiria, Leiria, Portugal

ESTG, Instituto Politécnico de Leiria, Leiria, Portugal

Jonathan Rodriguez Instituto de Telecomunicações, Aveiro, Portugal

Firooz B. Saghezchi Instituto de Telecomunicações, Aveiro, Portugal

Asunción Santamaría CeDInt-UPM. Campus de Montegancedo, Pozuelo de Alarcón, Madrid, Spain

Lambros Sarakis Electrical Engineering Department, Technical Institute of Sterea Ellada Psachna, Psachna, Greece

Luís Ducla Soares ISCTE - Instituto Universitário de Lisboa/Instituto de Telecomunicacões, Lisbon, Portugal

Nicolas Tizon VITEC Multimedia, Châtillon, France

Theodore Zahariadis Synelixis Solutions Ltd, Chalkida, Greece

Electrical Engineering Department, Technical Institute of Sterea Ellada Psachna, Psachna, Greece

Chapter 1
Introduction

Ahmet Kondoz and Tasos Dagiuklas

A significant amount of effort has been made by the research community towards achieving immersive multi-view entertainment. Most of the effort has been targeted at content acquisition and rendering/display technologies, or at delivery of pre-encoded content rather than live content. A key feature of immersive communication is the ability for participants to interact with the remote environment, and to detect and evaluate that interaction via their senses [1]. Thus, immersive communication is evolving in a variety of directions. One possible goal is to emulate reality, in which case high fidelity, realistic renderings are important and spatial relationships must be maintained. Another possible goal is to achieve a state of immersive communication that goes beyond reality, for example, enabling millions of people to attend a lecture or a music concert in which everyone has a front row seat [1, 2].

Commercial world has an increasing interest over 3D media. This is due to the fact that 3D media is viewed by the end user with glasses at home or cinemas. On the other hand, 3D media at commercial site can be considered in the following areas; delivery via broadcasting, via Video on Demand (VoD) and via storage devices such as DVDs. Currently commercial solutions for display side use either active or passive glasses for end users to experience 3D content. The content is delivered either in a single frame with left and right resolution is reduced to half (side-by-side or top-bottom methods), or left and right frames are packed in a single frame with full resolution (frame-packing).

A. Kondoz
Institute for Digital Technologies, Loughborough University in London, Queen Elizabeth Olympic Park, London E20 3ZH, UK
e-mail: a.kondoz@lboro.ac.uk

T. Dagiuklas (✉)
School of Science and Technology, Hellenic Open University, Parodos Aristotelous 18, Patras 26335, Greece
e-mail: dagiuklas@eap.gr

© Springer Science+Business Media New York 2015
A. Kondoz, T. Dagiuklas (eds.), *Novel 3D Media Technologies*,
DOI 10.1007/978-1-4939-2026-6_1

There are several EU projects on 3D media processing, delivery and presentation to the end-user. Amongst other, such projects include SEA, 2020 3D MEDIA, 3D4YOU, P2P-Next, MUSCADE, MOBILE3DTV, DIOMEDES, SKYMEDIA and ROMEO [3]. All these projects cover various issues ranging from 3D context creation to context-aware heterogeneous networking delivery platforms and QoE.

We are moving to a new era of convergence between media consumer and content provider. This shift is due to the new methods of media consumption via sharing using technologies such as converged heterogeneous networks, new transport methods, personal/user generated tools and social media as well as new multiple interactions and collaboration among the end-users [4]. As our lifestyles become more connected, even the passive behaviour of watching television is turning into a very active process, where viewers are multitasking on their mobile devices while watching new forms of content. This shift poses new challenges in jointly optimising media networking and sharing personalised content. Social TV refers to the convergence among broadcasting, media networking and social networking [5]. This gives the capability to the end-users to communicate, publish and share content among themselves and interact with the TV program. Consumers have aggressively adopted online video services (e.g. Netflix, YouTube, etc.). As more providers, more content, and more devices become available, consumers seem ready to take full advantage. More consumers also expect to see increased use of video on laptops, tablets and smartphones than on any other devices.

Looking from users' perspective, uploading their own content while enjoying the commonly shared content, as well as interacting either with the content which is being displayed or with their colleagues in the socially collaborating group is an important parameter to make the whole social interaction truly enjoyable [6]. A user on a mobile device (e.g. PDA, Tablet etc.) should be able to access the content from the closest point (distance, signal strength, availability) where the service is available. This particular service point is expected to also provide media processing capabilities to match the needs of the mobile terminal by pre-processing the available content in accordance to the given device capabilities, e.g. processing capability, resolution, content format it supports, etc. This could be supported by media storage and processing clouds that either the users will subscribe to (as an additional service) or their mobile network operator will provide it (within the subscription package) for increased user QoE (Quality of Experience) and hence increase the network's customer base.

Media processing in the cloud requires attention from two main perspectives: maintenance of processing related cloud operations over the execution time considering the end-user and application related QoS/QoE requirements via dynamic resource provisioning; and the parallelisation and abstraction of media processing tasks for the optimisation of the limited and shared cloud resources [7]. As an economic and scalable solution of enabling heterogeneous devices of various capabilities to handle media-related processing tasks, cloud-based media processing has gained lots of interest in the recent years. Within a cloud, multiple independent or inter-dependent media processing tasks could be dynamically spawn to run in parallel and share resources with each other in the cloud. Especially considering the extent to which the number of such sub-media processing tasks can reach in line with the applications as well as the user population, in such shared environments, how to appropriately provision computing resources to different tasks dynamically with regards to both system efficiency and Quality of Service remains a challenging problem. Different processing tasks, such as core multimedia encoding standards (e.g. H.264, HEVC, etc.) and computational tasks (e.g. Motion Estimation/Motion Compensation, DCT calculation, etc.), packetisation, transcoding, retargeting including transform domain and pixel domain processing, rendering (e.g. free-viewpoint video rendering), adaptation and mash-up with other content sources (e.g. web applications or other User Generated Content) have different computational, memory and bandwidth intensities. Most of the existing solutions focus only on maximising the system efficiency considering generic tasks, without necessarily overseeing the QoS requirements of various media processing sub-tasks and in overall the QoE offered to the end user. In that sense, the deployment of dynamic resource allocation within cloud architectures has been favoured to static resource allocation. Machine learning-based resource allocation for media processing tasks aims at getting a more precise prediction for various multimedia processing related sub-tasks based on historical data (e.g. average memory footprint, variance over time for various applications contexts, etc.).

This book describes recent innovations in 3D media and technologies, with coverage of 3D media capturing, processing, encoding, and adaptation, networking aspects for 3D Media, and quality of user experience (QoE). The structure of the book is the following:

Chapter 2 is entitled "Novel approaches to Immersive Media: from enlarged field-of-view to multi-sensorial experiences". The chapter presents a review of current evidence on the influence of immersion (defined in terms of the technical features of the system) on the user experience in multimedia applications. The chapter introduces the concepts of media enjoyment, presence and Quality of Experience (QoE) that frame the analysis from the user perspective. It discusses the bounding effects of multimodal perception on the previously defined metrics. Furthermore, it analyses the influence of relevant technical factors on presence, enjoyment and QoE, with emphasis on those characterising the level of immersion delivered by system across four dimensions: inclusiveness, extensiveness, surrounding and vividness. Finally, it presents recent works integrating some of these factors into multi-sensorial media experiences and highlights open issues and research challenges to be tackled in order to deliver cost-effective multi-sensorial media solutions to the mass market.

Chapter 3 is entitled "3D video representation and coding". The technologies which allow an immersive user experience in 3D environments are rapidly evolving and new services have emerged. Most of these services require the use of 3D video, combined with appropriate display systems. As a consequence, research and development in 3D video continues attracting sustained interest. While stereoscopic viewing is already widely spread, namely in TV and gaming, new displays and applications, such as FTV (Free viewpoint TV), require the use of a larger number of views. Hence, the multi-view video format was considered, which uses N views, corresponding to the images captured by N cameras (either real or virtual), with a controlled spatial arranged. In order to avoid a linear escalation of the bitrate, associated with the use of multiple views, video-plus-depth formats have been proposed. A small number of colour and depth video sequences are used to synthesise intermediate colour views at a different space position, through a depth-image-based rendering (DIBR) technique. This technology allows the use of advanced stereoscopic display processing and to improve support for high-quality auto-stereoscopic multi-view displays. In order to provide a true 3D content and fatigue-free 3D visualisation, holoscopic imaging has been introduced as an acquisition and display solution. However, efficient coding schemes for this particular type of content are needed to enable proper storage and delivery of the large amount of data involved in these systems, which is also addressed in this chapter.

Chapter 4 is entitled "Super Multi-View video". It presents motion parallax which is a key cue in the perception of the depth that current 3D stereoscopy and auto-stereoscopy technologies are not able to reproduce. Integral imaging and Super Multi-View video (SMV) are 3D technologies that allow creating a light-field representation of a scene with a smooth full motion parallax (i.e. in horizontal and vertical directions). However the large amount of data required is challenging and implies the need for new efficient coding technologies. This chapter first describes integral imaging and SMV content, acquisition and display. Then it provides an overview of state of the art methods for full parallax 3D content compression. Finally, several coding schemes are compared and a coding structure that exploits inter-view correlations in both horizontal and vertical directions is presented. The new structure provides a rate reduction (for the same quality) up to 29.1 % when compared to a basic anchor structure. Neighbouring Block Disparity Vector (NBDV) and Inter-View Motion Prediction (IVMP) coding tools are further improved to efficiently exploit coding structures in two dimensions, with rate reduction up to 4.2 % with respect to the reference 3D-HEVC encoder.

Chapter 5 is entitled "3D Holoscopic Video Representation and Coding Technology". 3D Holoscopic Imaging, also known as integral imaging, light-field imaging or plenoptic imaging, has been attracting the attention of the research community as a prospective technology for 3D acquisition and visualisation. However, to make this technology a commercially viable candidate for three-dimensional services, there are several important requirements that still need to be fulfilled. This chapter provides an overview of some of these critical success factors with special emphasis on the need for a suitable representation format in conjunction with an efficient coding scheme for 3D holoscopic content. Moreover, an efficient 3D holoscopic coding solution is described, which provides backward compatibility with legacy 2D and 3D multi-view displays by using a multi-layer scalable coding architecture. In addition to this, novel prediction methods are presented to enable efficient storage and delivery of 3D holoscopic content to the end-users.

Chapter 6 is entitled "Visual Attention Modelling in a 3D Context". This chapter provides a general framework for visual attention modelling. A combination of different state-of-the-art approaches in the field of saliency detection is presented.

This is accomplished by extending various spatial domain approaches to the temporal domain. Proposed saliency detection methods (with and without using depth information) are applied on the video to detect salient regions. Finally, experimental results are shown in order to validate the saliency map quality with the eye tracking system results. This chapter also deals with the integration of visual attention models in video compression algorithms. Jointly with the eye tracking data, this use case provides a validation framework to assess the relevance of saliency extraction methods.

Chapter 7 is entitled "Dynamic cloud resource migration for efficient 3D video processing in mobile computing environments". The chapter presents a dynamic cloud computing scheme for efficient resource migration and 3D media content processing in mobile computing environments. It elaborates on location and capacity issues to offload resources from mobile devices due to their processing limitations, towards efficiently manipulating 3D video content. The proposed scheme adopts a rack-based approach that enables cooperative migration for redundant resources to be offloaded towards facilitating 3D media content manipulation. The rack-based approach significantly reduces crash failures that lead all servers to become unavailable within a rack and enables mobile devices with limited processing capabilities to reproduce multimedia services at an acceptable level of Quality of Experience (QoE). The presented scheme is thoroughly evaluated through simulation tests, where the resource migration policy was used in the context of cloud rack failures for delay-bounded resource availability of mobile users.

Chapter 8 is entitled "Cooperative Strategies for End-to-End Energy Saving and QoS Control". Energy efficiency and Quality of Service (QoS) have become major requirements in the research and commercial community to develop green communication technologies for cost-effective and seamless convergence of all services (e.g. data, 3D media, Haptics, etc.) over the Internet. In particular, efforts in wireless networks demonstrate that energy saving can be achieved through cooperative communication techniques such as multihop communications or cooperative relaying. Game-theoretic techniques are known to enable interactions between collaborating entities in which each player can dynamically adopt a strategy that maximises the number of bits successfully transmitted per unit of energy consumed, contributing to the overall optimisation of the network in a distributed fashion. As for the core networks, recent findings claimed that resource over-provisioning is promising since it allows for dynamically booking more resources in advance, and multiple service requests can be admitted in a network without incurring the traditional per-flow signalling, while guaranteeing differentiated QoS. Indeed, heavy control signalling load has recently raised unprecedented concerns due to the related undue processing overhead in terms of energy, CPU, and memory consumption. While cooperative communication and resource over-provisioning have been researched for many years, the former focuses on wireless access and the latter on the wired core networks only. Therefore, this chapter investigates existing solutions in these fields and proposes new design approach and guidelines to integrate both access and core technologies in such a way as to provide a scalable and energy-efficient support for end-to-end QoS-enabled communication control. This is of paramount importance to ensure rapid development of attractive 3D media streaming in the current and future Internet.

Chapter 9 is entitled "Real-Time 3D QoE Evaluation of Novel 3D Media". Recent wireless networks enable the transmission of high bandwidth multimedia data, including advanced 3D video applications. Measuring 3D video quality is a challenge task due to a number of perceptual attributes associated with 3D video viewing (e.g. image quality, depth perception, naturalness). Subjective as well as objective metrics have been developed to measure 3D video quality against different artefacts. However most of these metrics are Full-Reference (FR) quality metrics and require the original 3D video sequence to measure the quality at the receiver-end. Therefore, these are not a viable solution for system monitoring/update "on the fly". This chapter presents a Near No-Reference (NR) quality metric for colour plus depth 3D video compression and transmission using the extracted edge information of colour images and depth maps. This work is motivated by the fact that the edges/contours of the depth map and of the corresponding colour image can represent different depth levels and identify image objects/boundaries of the corresponding colour image and hence can be used in quality evaluation. The performance of the proposed method is evaluated for different compression ratios and network conditions. The results obtained match well those achieved with its counterpart FR quality metric and with subjective tests, with only a few bytes of overhead for the original 3D image sequence as side-information.

Chapter 10 is entitled "Visual discomfort in 3DTV–Definitions, causes, measurement, and modeling". This chapter discusses the phenomenon of visual discomfort in stereoscopic and multi-view 3D video reproduction. Distinctive definitions are provided for visual discomfort and visual fatigue. The sources of visual discomfort are evaluated in a qualitative and quantitative manner. Various technological influence factors, such as camera shootings, displays and viewing conditions are considered, providing numerical limits for technical parameters when appropriate and available. Visual discomfort is strongly related to the displayed content and its properties, notably the spatiotemporal disparity distribution. Characterising the influence of content properties requires well-controlled subjective experiments and a rigorous statistical analysis.

Chapter 11 is entitled "3D Sound Reproduction by Wave Field Synthesis". This chapter gives emphasis on a spatial audio technique implemented for 3D multimedia content. The spatial audio rendering method based on wave field synthesis is particularly useful for applications where multiple listeners experience a true spatial soundscape while being free to move around without losing spatial sound properties. The approach can be considered as a general solution to the static listening restriction imposed by conventional methods, which rely on an accurate sound reproduction within a sweet spot only. While covering the majority of the listening area, the approach based on wave field synthesis can create a variety of virtual audio objects at target positions with very high accuracy. An accurate spatial impression could be achieved by wave field synthesis with multiple simultaneous audible depth cues improving localisation accuracy over single object rendering.

The current difficulties and practical limitations of the method are also discussed and clarified in this chapter.

Chapter 12 is entitled "Utilizing social interaction information for efficient 3D immersive overlay communications". Tele-immersive 3D communications pose significant challenges in networking research and request for efficient construction of overlay networks, to guarantee the efficient delivery. In the last couple of years, various overlay construction methods have been subject to many research projects and studies. However, in most cases, the selection of the overlay nodes is mainly based on network and geographic only criteria. In this book chapter, we focus on the social interaction of the participants and the way that social interaction could be taken into account for the construction of a network multicasting overlay. More precisely, we mine information from the social network structures and correlate it with network characteristics to select the nodes that could potentially serve more than one users, and thus contribute towards the overall network overlay optimisation.

References

1. Apostolopoulos J et al (2002) The road to immersive communication. Proc IEEE 100 (4):974–990
2. Chou P (2013) Advances in immersive communication: (1) Telephone, (2) Television, (3) Teleportation. ACM Trans Multimedia Computing, Commun Appl 9(1):41–44
3. Ekmekcioglu E et al (2012) Immersive 3D media. In: Moustafa H, Zeadally S (eds) Media networks: architect, Appl Standards. CRC Press
4. Montpetit M-J, Klym N, Mirlacher T (2011) The future of IPTV. Multimed Tools Appl 53 (3):519–532
5. Chorianopoulos K, Lekakos G (2008) Introduction to social TV: enhancing the shared experience with interactive TV. Int J Hum–Comput Int 24(2):113–120
6. Zhu W, Luo C, Wang J, Li S (2011) Multimedia cloud computing. IEEE Signal Proc Mag 28:50–61
7. Wen Y et al (2014) Cloud mobile media: reflections and outlook. IEEE Trans Multimedia 16 (4):885–902

Chapter 2
Novel Approaches to Immersive Media: From Enlarged Field-of-View to Multi-sensorial Experiences

Iris Galloso, Claudio Feijóo, and Asunción Santamaría

Abstract This chapter presents a review of current evidence on the influence of immersion (defined in terms of the technical features of the system) on the user experience in multimedia applications. Section 2.1 introduces the concepts of media enjoyment, presence, and Quality of Experience (QoE) that frame our analysis from the user perspective. Section 2.2 discusses the bounding effects of multimodal perception on the previously defined metrics. Section 2.3 analyses the influence of relevant technical factors on presence, enjoyment, and QoE, with emphasis on those characterizing the level of immersion delivered by system across four dimensions: inclusiveness, extensiveness, surrounding, and vividness. Section 2.4 presents recent works integrating some of these factors into multi-sensorial media experiences and highlights open issues and research challenges to be tackled in order to deliver cost-effective multi-sensorial media solutions to the mass market.

2.1 Conceptualizing User Experience with Entertaining Technologies

2.1.1 Media Enjoyment

Consistent results across more than seven decades of mass media effects research (in particular, under the uses and gratifications approach) identify enjoyment as the primary gratification sought from media [1]. Considered a direct predictor of audience, **media enjoyment** has been in the focus of media effects research for almost 40 years [1, 2].

I. Galloso (✉) • C. Feijóo • A. Santamaría
CeDInt-UPM. Campus de Montegancedo, Pozuelo de Alarcón 28223, Madrid, Spain
e-mail: iris@cedint.upm.es; cfeijoo@cedint.upm.es; asun@cedint.upm.es

© Springer Science+Business Media New York 2015
A. Kondoz, T. Dagiuklas (eds.), *Novel 3D Media Technologies*,
DOI 10.1007/978-1-4939-2026-6_2

The encyclopedia of positive psychology defines enjoyment as *"engagement in a challenging experience that either includes or results in a positive affective state"* [3]. Csikszentmihalyi, the father of the theory of the optimal experience, conceptualizes the term beyond pleasure, arguing that enjoyment is characterized by *"forward movement that accomplishes something novel or challenging, resulting in a growth experience"* [4]. Indeed, an enjoyable media experience presents several features inherent to the state of flow, such as: intense and focused concentration, merging of action and awareness, loss of reflective self-consciousness, distortion of temporal experience, and experience of the activity as intrinsically rewarding [5].

The components and dynamics underlying media enjoyment have been studied across a great variety of genres as a dependent variable of personality traits, individual differences, mood, content characteristics, social context, or a combination of these. As a result, it has been characterized as a multidimensional construct conditioned by affective components in a first place but also by cognitive and behavioral factors [6–11]. In particular, emotional enjoyment has been found closely linked to entertainment as a media effect, which at the same time correlates with some of the more frequently reported motivations for media use: arousal, to pass time, relaxation, and to escape. In this sense, media provides a mean to *"escape to a fantasy world where emotions can be experienced"* [1].

2.1.2 Presence

The desire of escaping from reality (or to some extent, of being "transported" to a different place) leads to the concept of **presence**, which is defined as *"the subjective experience of being in one place"*, even when the person is physically located in another [12].

The sense of presence has been widely analyzed as a mean to describe the psychological mechanisms underlying user experience with entertaining technologies, with particular emphasis on interactive computer-generated applications. Presence has been found strongly related to the capability of mediated environments—including 3DTV, videogames, and artistic and cultural heritage Virtual Environments (VEs)—to elicit emotions [13, 14] and in particular, enjoyment [15, 16]. In consequence, an enhanced sense of presence is considered to have a direct impact on the adoption potential of these entertaining applications.

The factors influencing the subjective sense of presence can be classified as those related to the media form, the media content and the media users [17]. Media form is related to the *extent and fidelity of sensory information* and to the consistency between the sensory inputs/outputs presented through the different modalities [12, 17, 18]. In other words, it encompasses those features characterizing in an objective manner the capability of a specific software and hardware solution to deliver a rich and consistent multi-sensorial media content in a transparent manner (i.e., as an invisible medium). The influence of relevant media form and content factors in the sense of presence is analyzed in Sect. 2.3.

Media content is a very broad category concerning issues as the story, messages, objects, activities, or characters presented as part of the dynamic experience. Among the content characteristics identified as determinants of presence are: social realism, representation of virtual body, autonomous behavior and appearance of characters and objects, the ability to modify the physical environment, and to anticipate the effect of an action and possible interactions between the type, nature, and complexity of tasks or activities [17–20].

As regards to the characteristics of the media user, the sense of presence has been found significantly influenced by emotional, cognitive, and motivational-behavioral factors, such as: immersive tendency (measured in terms of absorption and emotional involvement, which at the same time correlate with openness to experience and with neuroticism and extraversion in the last case), attention, relevance, skill, perceived control, challenge, cognitive capabilities (e.g., spatial intelligence), and personality traits [21–26]. In particular, works as [24] provide evidence on the impact of user features as competence, challenge, and ability to explore the VE as well as of media form variables as interaction speed, mapping, and range on the spatial presence. The study also supports previous findings on the relation between the emotional response (in terms of arousal) and the levels of attention, spatial presence, and situational involvement. Such results point to a significant influence of the individual's cognitive–affective assessment of the immersive media form and content on the emotional response and the perceived level of physical presence.

2.1.3 Quality of Experience

The factors and mechanisms that influence the subjective quality assessment of a multimedia asset (i.e., the content quality as perceived by an individual) are analyzed by researchers in the field of user experience in multimedia applications. The study of these phenomena has been encompassed into the concept of "**Quality of Experience (QoE)**," which is defined as *"the degree of delight or annoyance of the user of an application or service (...) which results from the fulfillment of [his/her] expectations with respect to the utility and/or enjoyment of the application or service in the light of the his/her personality and current state"* [27]. In this context, the user experience has been found influenced by a combination of interrelated factors of contextual, technical, and human nature.

Contextual factors have been defined as those *"that embrace any situational property to describe the user's environment"* [28]. These not only concern the physical context, but also other dynamic or static features of economic, social, or technical nature [27]. Research on the influence of contextual factors is out of the scope of this chapter.

Technical factors (also known as system factors) refer to those conditioning the resulting technical quality of an application or service [28]. Different categories of system factors have been proposed in literature, both from a technical perspective, in which they are divided according to the related components of the service/ architecture chain and from a user perspective, considering their final influence/ manifestation during the experience [29]. Relevant findings on the influence of system factors on the QoE are analyzed in Sect. 2.3.

Human factors comprise those features that characterize the user and have an influence on his/her perception of quality. Quality perception is framed by the human perception mechanism, which flows at two main levels: the early sensory processing level, aimed at extracting relevant features from the incoming multi-modal sensory information, and the high-level cognitive processing level, focused on conscious interpretation and judgment [28, 30]. This dynamic is supported from a psychological perspective by Lazarus' theory of appraisal [31]; in which primary appraisal involves an appraisal of the outside world and secondary appraisal involves an appraisal of the individual themselves.

Although this classification has been useful for analysis purposes, the boundaries between the two processing levels are not clearly established. In contrast, there is strong evidence pointing to a modulating effect of high level factors as knowledge, emotions, expectations, attitudes, and goals on the relative importance of sensory modalities and their attributes, as well as on the orientation of attentional resources accordingly [32–34]. These changes in early sensory processing can be subject to a specific domain of expertise (e.g., image-based diagnosis) [35, 36] or can be eventually consolidated as a general ability [37, 38]. Furthermore, in case of discrepancies between the individual knowledge schema (built from past experiences and from abstract expectations and representations of the external reality) and the sensory input, the structure of the schema can be modified to integrate the contradictory stimuli (i.e., absorption of new knowledge) [39].

2.2   The Bounding Effect of Multimodal Perception

An extended belief in the presence research community is that the more extensive an immersive system is (i.e., in terms of its capability to stimulate a greater number of human senses), the greater its capability to evoke presence (see [17] and citations thereof). This hypothesis is supported by works as [40], where the addition of tactile, olfactory, and auditory cues showed to have a positive impact on the sense of presence. Likewise, in [41] an inverse correlation between the mental processing times (i.e., simple detection times) and the number of sensory modalities presented (unimodal, bimodal, or trimodal) was found. However, these results can't be generalized in a straightforward manner to the quality perception context.

Multimodal perception is a complex phenomenon dealing with the integration of two or more sensory inputs into a single, coherent, and meaningful stimulus. Although the factors influencing the perceptual experience have not been entirely characterized yet, there is strong evidence on the integration and sharing of perceptual information since the very early sensory processing stages [42, 43] and on the bounding effects of cognitive, emotional, and personality factors [10].

The presence of a given modality can distort or modulate (either intensifying or attenuating) the perception in other modality. Cross-modal interaction processes— as for example, synesthesia—underline the relative importance of the different sensory modalities presented (see [43] and citations thereof). This phenomenon has been widely analyzed, from an empirical perspective, in terms of the relative influence of vision and sound on task-related performance [44–48]. Findings reveal the potential of vision to alter the perception of speech and spatial location of audio sources and the influence of audio on vision in terms of temporal resolution, intensity, quality, structure, and interpretation of visual motion events. Concerning other modalities, a form of synesthesia—defined as "crossmodal transfer"—has been reported between vision and touch. The phenomenon is characterized by the appearance of a perceptual illusion in one modality induced by a correlated stimulus on other sensory modality (e.g., illusion of physical resistance induced by the manipulation of a virtual object in a mediated environment) [49]. Interestingly (although not surprisingly), this cross-modal illusion was found correlated with the sensation of spatial and sensory presence in the displayed environment. In [41], participants reacted faster (i.e., lower simple detection times were measured) to auditory and haptic stimulus than to visual stimulus when only one of them was presented (unimodal condition). In coherence, the bimodal auditory–haptic combination resulted in even faster reactions in comparison to those reported for each unimodal component and for the other two bimodal combinations (visual-haptic and visual-auditory). These results suggest a highly relevant influence of auditory and haptic stimuli on processing times at the initial perceptual stage, which according to the authors allows users more time in the consequent cognitive stages, enabling them better integration and filling in of missing information. Similarly, the authors in [50] found that haptic feedback can led to an improved task performance and feeling of "sense of togetherness" in shared VEs.

The majority of these empirical findings support the "modality appropriateness" hypothesis, which argues that *the modality that is most appropriate or reliable with respect to a given task dominates the perception in the context of that task* [51]. However, this and other approaches still require further elaboration to better explain complex effects as the wide variety of responses to inter-sensory divergent events reported in literature.

2.3 The Influence of Immersion

The effectiveness of a mediated environment to evoke cognitive and emotional reactions in a similar way to non-mediated experiences is heavily conditioned by the consistency between the displayed environment and an equivalent real environment as regards to the user experience [52, 53]. Two main components contributing to this realistic response are identified in [54]. These are: *place illusion*, defined as *the subjective sensation of being in a real place* (i.e., presence); and, *plausibility illusion*, referred as the *illusion that the scenario being depicted is actually occurring*, even when the person is cognitively aware of the fact that it isn't. In this sense, the plausibility judgment is highly related to the capability of the system as a whole to produce events that are meaningful and credible in comparison to the individual's expectations [54].

The capability of a technical system "*to deliver an inclusive, extensive, surrounding and vivid illusion of reality to the senses of a human participant*" has been defined as *immersion* [12, 55]. At this point, it should be emphasized a conceptual difference observed along this chapter between *immersion*, describing the capabilities of the system in an objective manner, and *presence*, considered a state of consciousness derived from the subjective perception of the experience [55].

An immersive system can be characterized in terms of four major dimensions as: *inclusive*, the extent to which it is able to isolate the physical reality; *extensive*, the range of sensory modalities addressed; *surrounding*, the extent to which the user is physically surrounded by the displayed environment; and, *vivid*, the resolution, fidelity, and variety of the sensorial stimuli delivered through each sensory modality. Each of these dimensions can be present at different levels and scales according to the correlating psycho–physiological responses and to the extent of their realization, respectively [12, 56].

2.3.1 Breakdown of System Factors

The independent and combined influence of system factors (including media form and content variables) on the emotional response, on the subjective assessment of presence and on quality judgment (in terms of QoE) has been analyzed extensively in scientific literature. In this section, we present and discuss relevant findings illustrating the complexity and wide variety of approaches to these fields on a non-exhaustive basis. Table 2.1 summarizes in a schematic way the facts analyzed as follows.

The influence of factors such as image motion, stereoscopy, and screen size has been studied in [57]. Image motion and stereoscopy showed to have, in that order, a great influence on presence. A large effect of screen size on presence was also observed, but only for the video stimulus that contained motion. High motion content has also shown an impact on the relative quality of video and audio perceived by the user, being the video quality weighted significantly higher than the audio quality when high motion content is presented [58].

A relationship between motion-based interaction and the perceived field-of-view (FOV) is reported in [59]. The perceived FOV for a small-hand held device was found around 50 % greater than the actual value when motion-based interaction was used. Coherently, the sense of presence under this condition was higher than or comparable to those in VR platforms with larger displays. The effects of head tracking, visual cues (including stereoscopic and motion parallax cues), and

Table 2.1 Influence of system factors on the emotional response, the sense of presence, and the QoE

System factors	Influence on the perceived... (correlation sign in parenthesis)	Sensory modality(ies) addressed	Main dimension (s) involved
Image motion [57, 58]	Presence (+), relative quality of video and audio (trade-off)	Sight	Vividness
Interaction between image motion and screen size [57]	Presence (+)	Sight	Vividness
Stereoscopy and stereoscopic and motion parallax cues [57, 60, 62]	Presence (+)	Sight	Vividness
Visual cues (spatial and object cues) [22]	Presence (+)	Sight	Vividness
Screen size, geometric field-of-view, omnidirectional video [57, 59–61, 63]	Presence (+), simulator sickness (+), enjoyment (−)	Sight	Inclusiveness, surrounding
Interaction between motion-based interaction and perceived field-of-view [59]	Presence (+)	Sight, proprioception	Extensiveness, inclusiveness
Interaction between natural (hand-based) interaction and narrative [63]	Presence (trade-off)	Sight, proprioception	Extensiveness
Pictorial realism [64, 65]	Presence (+)	Sight	Vividness
Delay of visual feedback [12, 65]	Presence (−)	Sight	Vividness
Presence or absence of interactivity [65]	Presence (+)	Sight, touch	Extensiveness
Frame rate [66, 67]	Presence (+)	Sight	Vividness
Passive haptic feedback [67]	Presence (+)	Sight, touch	Extensiveness
Presence or absence of spatialized sound, addition of spatialized versus non-spatialized sound to a stereoscopic display [14, 60, 68]	Presence (+), QoE (+), emotional response (+), emotion recognition (+), and emotional realism (+)	Hearing	Vividness
Image quality [69]	Audio quality (+), audiovisual quality (+)	Sight	Vividness
Audio quality [48, 69]	Presence (+), enjoyment (+), visual quality (+)	Hearing	Vividness
Natural physical interactions: head tracking [60], walk in place [70]	Presence (+)	Sight, touch, proprioception, equilibrioception	Extensiveness
Sensory effects (wind, vibration, light effects) and genre [71, 72]	QoE (+), enjoyment (+)	Sight, touch, thermoception	Extensiveness, surrounding, inclusiveness

(continued)

Table 2.1 (continued)

System factors	Influence on the perceived... (correlation sign in parenthesis)	Sensory modality(ies) addressed	Main dimension (s) involved
Olfactory effects [73–75]	QoE (+), relevance (+), reality (+), and enjoyment (+)	Olfacception	Extensiveness, surrounding, inclusiveness
Synchronization errors (outside the tolerance range) between video + audio or video without audio and olfaction [73, 74]	QoE (−), relevance (−), reality (−), and enjoyment (−)	Sight, hearing, olfacception	Vividness
Audio–video asynchrony (in particular, audio-led asynchrony) [28, 77, 78]	Clarity of the message (−), distraction (+)	Sight, hearing	Vividness
Stereoscopic disparities: large disparity at short convergence distances [79, 81]	Presence (−), enjoyment (−), QoE (−)	Sight	Vividness
(In stereoscopy) spatial distortions: shifts, magnification, rotation, keystone [80, 81]	Presence (−), enjoyment (−), QoE (−)	Sight	Vividness
(In stereoscopy) photometric asymmetries: luminance, color, contrast, crosstalk [80, 81]	Presence (−), enjoyment (−), QoE (−)	Sight	Vividness
Immersive technology (PC, big screen, HMD) [82]	Simulation sickness (+)	Sight	Inclusiveness, surrounding

geometric field-of-view are also explored in [22, 60–62]. The reported level of presence was positively influenced by the use of tracking, stereoscopic, and spatial and object cues [22, 60]. Presence was also correlated with the geometric field-of-view, showing an asymptotic behavior for field-of-view values beyond 140° [60, 61].

The experience of a theatrical performance and television using interactive omnidirectional video is qualitatively explored in [63]. Participants referred to the experience—in cognitive and physical terms—as being *discovering and exploring the (mediated) environment*. They also described transitions between the real and the displayed environment as disturbing and therefore, requiring a recalibration of the senses. Under this engaging experience, narrative was pushed to a second place and the hand-based interaction put in place was qualified as highly intuitive. The authors conclude that interactivity may influence the perception of narrative and therefore, these factors need to be carefully balanced to maximize presence.

Pictorial realism, observer interactivity, and delay of visual feedback are analyzed in [64, 65]. Realism and interactivity were shown to have a positive impact on presence while delay of visual feedback had an opposite effect. Participants reported a relative low influence of pictorial realism on presence in comparison to the other two components considered. The influence on presence of other screen variables as the frame rate has been shown in works as [66, 67].

As regards to the influence of auditory features, the audio quality, a realistic aural rendering of events, and the presence of auditory cues are considered to have a significant impact on the sense of presence [14, 48, 60]. The influence of realistic aural rendering, measured in terms of the number of audio channels (mono, stereo, and six-channel reproduction), on presence, emotional response and emotion recognition is analyzed in [14]. Stereo and six-channel reproduction had a significantly stronger impact in emotional response than the mono condition. Similarly, six-channel reproduction resulted in the highest ratings of presence and emotional realism. In coherence, an enhanced sense of presence and QoE are reported in [60] and [68] respectively, in response to the addition of spatialized audio. In [48], the relative influence of image quality (high definition vs. standard definition) and sound quality (Dolby 5.1 surround sound vs. Dolby stereo) on presence and enjoyment is studied. No significant effects of image quality were found. In contrast, the impact of sound quality on presence and enjoyment was shown to be significant. Furthermore, a significant cross-modal influence of audio on visual quality and vice versa has been reported in [69], although video quality seemed to dominate the overall perceived audiovisual quality in the context of the study.

The introduction of interaction has been also found to be significant [65]. In particular, interactions entailing natural physical movements—e.g., head movement [60] or walking in place [70]—and leading to a coherent system response (as regards to the individual's expectations) have shown a great impact on presence. Likewise, a significant influence of passive haptic feedback on presence has been reported in [67].

Less traditional stimuli as wind, vibration, and light effects have also shown a significant impact on the user experience (both in terms of enjoyment [71] and QoE), in particular with genres as action movies, sports, news, documentary, and commercials [72]. Likewise, olfactory effects have shown a positive influence on the perceived quality, relevance, and reality and on the reported enjoyment of a multimedia experience [73–75]. A potential exception to these positive effects may be given by synchronization errors producing a mismatch between video + audio and olfaction that is outside the temporal range of −30s (olfaction ahead of video + audio) to +20s (video + audio ahead of olfaction) [73]. However, in the case of video without audio, the tolerance to synchronization errors with olfaction decreases [74].

Technological breakdowns significantly reduce the potential of mediated environments to elicit presence and emotions [76]. For instance, an asynchronous reproduction of audio and video in the context of an audiovisual experience has shown a negative impact on the clarity of the message, distracting the viewer from the intended content [28]. In particular, users are more sensitive and report higher annoying effects in the case of audio-led asynchrony [77, 78]. Concerning stereoscopy, the variables influencing visual comfort in a negative fashion can be classified as: those introducing spatial distortions as shifts, magnification, rotation, and keystone; those leading to photometric asymmetries as luminance, color, contrast, and crosstalk; and, those leading to stereoscopic disparities as the disparity level, which shows a larger effect at short convergence distances [79–81]. Other meaningful studies show, for instance, that the level of simulator sickness is positively correlated with the geometric field-of-view [61]. Interestingly, presence was positively correlated with simulator sickness while enjoyment showed the opposite behavior. Similarly, a relationship between the immersive technology used and the severity of the negative effects reported was found in [82]. From the three immersive technologies analyzed (PC, Head Mounted Display (HMD), and big screen), HMD was the one producing more negative effects.

2.4 Implementing Multi-sensorial Media: Current Issues and Future Challenges

In an attempt to deliver a more immersive experience (i.e., more extensive, inclusive, surrounding, and vivid and in consequence, more enjoyable), several works propose the integration of sensory effects (beyond the conventional audiovisual content) into a multimedia asset. In particular, the concept seems to have the potential to bring actual immersive experiences to the home in a non-disruptive manner. That is, presenting sensory effects as a complement to current display technology that can be progressively adopted in transparent way.

An early initiative introducing meaningful lighting effects as a mean to complement the main audiovisual content is illustrated in [71]. Using their HomeLab research facility, the authors installed the Philips Living Light system. The system comprised four LightSpeakers (left–right front–back), a CenterLight, and a SubLight (situated underneath the couch). Ad-hoc light scripts were developed, with the support of light designers, theatre lighting experts, filmmakers, and musicians, for selected pieces of film and music. In the qualitative interview conducted participants expressed that *lighting effects made watching movies or listening to music a very enjoyable and more immersive experience.* The concept was also found appealing for creating personalized ambiances at home in the context of other social or personal activities.

The authors in [83] present sensory effects as a new dimension contributing to the QoE. The sensory effects are defined by the Sensory Effect Metadata (SEM) which should accompany or be retrieved together with the media content. The media processing engine is responsible for playing the audiovisual content and the corresponding sensory effects in a synchronized manner, considering the capabilities of the rendering devices. In one of their experiments, the authors analyze the influence of wind, vibration, and light effects in the user experience across different genres [72]. They found that the QoE was positively influenced by the introduction of sensory effects in the case of documentary, sports, and action genres. A less noticeable but still positive influence was found for commercials. As future research, Timmerer et al. [83] outline the need to establish a quality/utility model for sensory experiences and to develop (semi-)automatic annotation techniques for the generation and integration of sensory effects into media assets.

In [68] an end-to-end solution integrating sensory effects and interactive components into a hybrid (internet-broadcast) 3DTV system is presented. In the experimental setup deployed the main audiovisual content (showing an extended report of a football match) is complemented with binaural audio, cut grass scent, ambient lighting effects, and main lighting and shutter controllers (immersion dimension), and with interactive 3D objects and meaningful content delivered through a second screen (interaction dimension). A combination of broadcast–broadband transmission mechanisms is implemented to transmit this complementary content. At the user's premises, the content is delivered using the private IP network that connects the receiver gateway with the visualization terminals and sensory devices. The resulting system is compatible with current transmission (DVB-T), coding (AVC), multiplexing (MPEG-2), signaling (DVB), and automation (KNX) standards.

The development and official release of the MPEG-V standard by the Moving Picture Expert Group (MPEG) (and in particular, of its Part 3—Sensory Information [84]) represents an important step in the consolidation of the sensory experience concept. The standard establishes the architecture and associated information representations for the interaction and interoperability between virtual worlds (i.e., multimedia content) and real worlds through various sensors and actuators. The Part 3 defines a set of sensory effects (e.g., light, temperature, wind, vibration, touch) and associated semantics to deliver multi-sensorial content in association with multimedia.

A recent Special Issue on MPEG-V, released on February 2003, gathers several contributions proposing end-to-end frameworks that implement the standard for the creation and delivery of sensory effects synchronized with audiovisual content. Three relevant examples are those provided in [85–87]. In [85] an authoring tool called SEVino is used for the generation of the SEM descriptions corresponding to the different sensory effects introduced. The annotated content can be delivered over various distribution channels and visualized in any MPEG-V-compliant device. The SEM descriptions enable sensory effects to be rendered on off-the-shelf hardware synchronized with the main audiovisual content, either in a stand-alone application or in a web browser. Concerning the user experience, the authors confirmed the hypotheses that sensory effects have a positive impact on the QoE and on the intensity of emotions like happiness or fun.

The framework presented in [86] delivers sensory effects for home theaters based on MPEG-V standard via the broadcast network. The paper discusses thoroughly the technical choices provided by the MPEG-V standard (and those adopted in the targeted implementation) for the description, encoding, synchronization, transport, adaptation, and rendering of sensory effects. The work in [87] also exploits the broadcasting network capabilities to deliver a haptic-enabled system based on the MPEG-V standard. The paper illustrates the data flow within the system, which comprises four main stages: the creation of haptic contents using the MPEG-V standard, their encoding/decoding using BIFS encoders/decoders, their transmission through the MPEG-4 framework, and the rendering of the haptic effects using the MPEG-V standard in the rendering stage.

Important challenges remain to deliver a cost-effective implementation of multi-sensorial media solutions. A major issue identified by several authors is the need to establish a quality/utility model for sensory experiences. At the content creation stage, the development of effective (semi)automatic video annotation tools is a common challenge to the majority of multi-sensorial media implementations reviewed. Semantic video analysis seems a suitable strategy to identify those relevant events that should trigger sensory effects and/or interactive actions. A significant challenge is posed also by the use of computer vision algorithms to recognize specific scene features, objects, elements, or characters as a way to boost the visualization of additional content (i.e., sensory effects) associated to the recognized element/character. However, the cost-intensity of these algorithms needs to be decreased to enhance their deployment feasibility.

Other issues that should be subject to further analysis and/or improvement are: the identification of more efficient encoding/decoding methods (in particular for large SEM), the configuration of suitable transport mechanisms and the effective management of the various types of delays introduced along the transmission chain.

At the receiver side, automatic techniques are required for enabling the discovery, feature detection, and remote configuration of sensory devices. Likewise, effective automatic mechanisms shall be developed to adapt sensory effects to the capabilities of the specific rendering devices available at the user's premises (e.g., specific protocol, resolution, or time constraints).

From a market perspective, sensory-enhanced video has the potential to support the development and deployment of immersive media services targeting the wide domestic segment. The high consumption of action movies, sports, and documentaries in this context might favor the adoption of these solutions.

References

1. Sherry JL (2004) Flow and media enjoyment. Commun Theory 14:328–347. doi:10.1111/j.1468-2885.2004.tb00318.x
2. Ruggiero TE (2000) Uses and gratifications theory in the 21st century. Mass Commun Soc 3:3–37. doi:10.1207/S15327825MCS0301_02
3. Kapsner JC (2009) The encyclopedia of positive psychology. doi: 10.1111/b.9781405161251.2009.x
4. Csikszentmihalyi M (1990) Flow: the psychology of optimal experience. Harper & Row, New York
5. Nakamura J, Csikszentmihalyi M (2003) The construction of meaning through vital engagement. In: Keyes CLM, Haidt J (eds) Flourishing Posit. Psychol. Life well-lived. American Psychological Association, Washington, pp 83–104
6. Nabi RL, Krcmar M (2004) Conceptualizing media enjoyment as attitude: implications for mass media effects research. Commun Theory 14:288–310. doi:10.1111/j.1468-2885.2004.tb00316.x

7. Slater MD (2003) Alienation, aggression, and sensation seeking as predictors of adolescent use of violent film, computer, and website content. J Commun 53:105–121. doi:10.1111/j.1460-2466.2003.tb03008.x

8. Zillmann D, Vorderer P (2000) Media entertainment: the psychology of its appeal. Lawrence Erlbaum Associates, Mahwah, p 282

9. Raney AA, Bryant J (2002) Moral judgment and crime drama: an integrated theory of enjoyment. J Commun 52:402–415. doi:10.1111/j.1460-2466.2002.tb02552.x

10. Wechsung I, Schulz M, Engelbrecht K-P et al (2011) All users are (not) equal – the influence of user characteristics on perceived quality, modality choice and performance. In: Kobayashi T, Delgado RL-C (eds) Proceedings of the paralinguistic information and its integration in spoken dialogue systems workshop. Springer, New York, pp 175–186

11. Tamborini R, Bowman ND, Eden A et al (2010) Defining media enjoyment as the satisfaction of intrinsic needs. J Commun 60:758–777. doi:10.1111/j.1460-2466.2010.01513.x

12. Slater M, Wilbur S (1997) A framework for immersive virtual environments (FIVE): speculations on the role of presence in virtual environments. Presence-Teleop Virtual Environ 6:603–616

13. Riva G, Mantovani F, Capideville CS et al (2007) Affective interactions using virtual reality: The link between presence and emotions. Cyberpsychology Behav 10:45–56. doi:10.1089/cpb.2006.9993

14. Västfjäll D (2003) The subjective sense of presence, emotion recognition, and experienced emotions in auditory virtual environments. Cyberpsychology Behav 6:181–188. doi:10.1089/109493103321640374

15. Sylaiou S, Mania K, Karoulis A, White M (2010) Exploring the relationship between presence and enjoyment in a virtual museum. Int J Hum Comput Stud 68:243–253

16. Skalski P, Tamborini R, Shelton A et al (2010) Mapping the road to fun: Natural video game controllers, presence, and game enjoyment. New Media Soc 13:224–242

17. Lombard M, Ditton T (1997) At the heart of it all: the concept of presence. J Comput Commun 3:0. doi: 10.1111/j.1083-6101.1997.tb00072.x

18. IJsselsteijn W (2000) Presence: concept, determinants, and measurement. In: Proc. SPIE. Spie. pp 520–529

19. Slater M, Steed A, McCarthy J, Maringelli F (1998) The influence of body movement on subjective presence in virtual environments. Hum Factors 40:469–477

20. Schuemie MJ, van der Straaten P, Krijn M, van der Mast CAPG (2001) Research on presence in virtual reality: a survey. Cyberpsychol Behav 4:183–201. doi:10.1089/109493101300117884

21. Darken RP, Bernatovich D, Lawson JP, Peterson B (1999) Quantitative measures of presence in virtual environments: the roles of attention and spatial comprehension. Cyberpsychol Behav 2:337–347

22. Lee S, Kim GJ (2008) Effects of visual cues and sustained attention on spatial presence in virtual environments based on spatial and object distinction. Interact Comput 20:491–502

23. Novak T, Hoffman D, Yung Y (2000) Measuring the customer experience in online environments: a structural modeling approach. Mark Sci 19:22–44

24. Takatalo J, Nyman G, Laaksonen L (2008) Components of human experience in virtual environments. Comput Human Behav 24:1–15. doi:10.1016/j.chb.2006.11.003

25. Weibel D, Wissmath B, Mast FW (2010) Immersion in mediated environments: the role of personality traits. Cyberpsychol Behav Soc Netw 13:251–256

26. Alsina-Jurnet I, Gutierrez-Maldonado J (2010) Influence of personality and individual abilities on the sense of presence experienced in anxiety triggering virtual environments. Int J Hum Comput Stud 68:788–801

27. Le Callet P, Möller S, Perkis A (2012) Qualinet white paper on definitions of quality of experience. In: European network on quality of experience in multimedia systems and services (COST Action IC 1003), Lausanne, Switzerland, Version 1.1, June 3, 2012

28. Jumisko-Pyykkö S (2011) User-centered quality of experience and its evaluation methods for mobile television. Tampere University of Technology
29. Bracken C, Pettey G, Wu M (2011) Telepresence and attention: secondary task reaction time and media form. In: Proc. Int. Soc. Presence
30. Goldstein EB (2010) Sensation and perception. p 496
31. Lazarus RS (1993) From psychological stress to the emotions: a history of changing outlooks. Annu Rev Psychol 44:1–21. doi:10.1146/annurev.ps.44.020193.000245
32. Bey C, McAdams S (2002) Schema-based processing in auditory scene analysis. Percept Psychophys 64:844–854
33. Jennings JR, Van der Molen MW, Van der Veen FM, Debski KB (2002) Influence of preparatory schema on the speed of responses to spatially compatible and incompatible stimuli. Psychophysiology 39:496–504
34. Cui LC (2003) Do experts and naive observers judge printing quality differently? In: Miyake Y, Rasmussen DR (eds) Electron. Imaging 2004. International Society for Optics and Photonics, pp 132–145
35. Werner S, Thies B (2000) Is "Change Blindness" attenuated by domain-specific expertise? An expert-novices comparison of change detection in football images. Vis Cogn 7:163–173. doi:10.1080/135062800394748
36. Sowden PT, Davies IR, Roling P (2000) Perceptual learning of the detection of features in X-ray images: a functional role for improvements in adults' visual sensitivity? J Exp Psychol Hum Percept Perform 26:379–390
37. Curran T, Gibson L, Horne JH et al (2009) Expert image analysts show enhanced visual processing in change detection. Psychon Bull Rev 16:390–397. doi:10.3758/PBR.16.2.390
38. Sowden PT, Rose D, Davies IRL (2002) Perceptual learning of luminance contrast detection: specific for spatial frequency and retinal location but not orientation. Vision Res 42:1249–1258. doi:10.1016/S0042-6989(02)00019-6
39. Neisser U (1976) Cognition and reality: principles and implications of cognitive psychology. p 230.
40. Dinh HQ, Walker N, Hodges LF, Kobayashi A (1999) Evaluating the importance of multi-sensory input on memory and the sense of presence in virtual environments. In: Proc. IEEE virtual real (Cat. No. 99CB36316). IEEE Comput. Soc., pp 222–228
41. Hecht D, Reiner M, Halevy G (2006) Multimodal virtual environments: response times, attention, and presence. Presence-Teleop Virtual Environ 15:515–523
42. Coen M (2001) Multimodal integration-a biological view. In: Proc. Fifteenth Int. Jt. Conf. Artif. Intell. Seattle, WA, pp 1417–1424
43. Shimojo S, Shams L (2001) Sensory modalities are not separate modalities: plasticity and interactions. Curr Opin Neurobiol 11:505–509
44. Mcgurk H, Macdonald J (1976) Hearing lips and seeing voices. Nature 264:746–748
45. Rock I, Victor J (1964) Vision and touch: an experimentally created conflict between the two senses. Science 143:594–596
46. Scheier C, Nijhawan R, Shimojo S (1999) Sound alters visual temporal resolution. Invest Ophthalmol Vis Sci 40:4169
47. Chandrasekaran C, Ghazanfar AA (2011) When what you see is not what you hear. Nat Neurosci 14:675–676. doi:10.1038/nn.2843
48. Skalski P, Whitbred R (2010) Image versus sound: a comparison of formal feature effects on presence and video game enjoyment. Psychology J 8:67–84, doi: Article
49. Biocca F, Kim J, Choi Y (2001) Visual touch in virtual environments: an exploratory study of presence, multimodal interfaces, and cross-modal sensory illusions. Presence-Teleop Virtual Environ 10:247–265
50. Basdogan C, Ho C, Srinivasan MA, Slater MEL (2001) An experimental study on the role of touch in shared virtual environments. ACM Trans Comput Interact 7:443–460
51. Welch RB, Warren DH (1980) Immediate perceptual response to intersensory discrepancy. Psychol Bull 88:638–667

52. De Kort YAW, Ijsselsteijn WA, Kooijman J, Schuurmans Y (2003) Virtual laboratories: comparability of real and virtual environments for environmental psychology. Presence-Teleop Virtual Environ 12:360–373
53. Bowman DA, McMahan RP (2007) Virtual reality: how much immersion is enough? Computer (Long Beach Calif) 40:36–43. doi:10.1109/MC.2007.257
54. Slater M (2009) Place illusion and plausibility can lead to realistic behaviour in immersive virtual environments. Philos Trans R Soc Lond B Biol Sci 364:3549–3557. doi:10.1098/rstb.2009.0138
55. Slater M (2003) A note on presence terminology. Presence-Connect, pp 1–5
56. Steuer J (1992) Defining virtual reality: dimensions determining telepresence. J Commun 42:73–93. doi:10.1111/j.1460-2466.1992.tb00812.x
57. Ijsselsteijn W, de Ridder H, Freeman J et al (2001) Effects of stereoscopic presentation, image motion, and screen size on subjective and objective corroborative measures of presence. Presence-Teleop Virtual Environ 10:298–311
58. Hands DS (2004) A basic multimedia quality model. IEEE Trans Multimedia 6:806–816. doi:10.1109/TMM.2004.837233
59. Hwang J, Kim GJ (2010) Provision and maintenance of presence and immersion in hand-held virtual reality through motion based interaction. Comput Animat Virtual Worlds 21:547–559
60. Hendrix C, Barfield W (1995) Presence in virtual environments as a function of visual and auditory cues. In: Proceedings of Virtual Real Annu Int Symp. pp 74–82. doi: 10.1109/VRAIS.1995.512482
61. Lin JJ-W, Duh HBL, Parker DE et al (2002) Effects of field of view on presence, enjoyment, memory, and simulator sickness in a virtual environment. In: Proc. IEEE Virtual Real. IEEE Comput. Soc., pp 164–171
62. Freeman J, Avons SE, Pearson DE, IJsselsteijn WA (1999) Effects of sensory information and prior experience on direct subjective ratings of presence. Presence-Teleop Virtual Environ 8:1–13. doi:10.1162/105474699566017
63. Bleumers L, Lievens B, Pierson J (2011) From sensory dream to television format: gathering user feedback on the use and experience of omnidirectional video-based solutions. In: ISPR 2011 Int. Soc. PRESENCE Res. Annu. Conf.
64. Slater M, Khanna P, Mortensen J, Yu I (2009) Visual realism enhances realistic response in an immersive virtual environment. IEEE Comput Graph Appl 29:76–84
65. Welch RB, Blackmon TT, Liu A et al (1996) The effects of pictorial realism, delay of visual feedback, and observer interactivity an the subjective sense of presence. Presence-Teleop Virtual Environ 5:263–273
66. Barfield W, Hendrix C (1995) The effect of update rate on the sense of presence within virtual environments. Virtual Real 1:3–15. doi:10.1007/BF02009709
67. Meehan M (2001) Physiological reaction as an objective measure of presence in virtual environments. University of North Carolina at Chapel Hill
68. Luque FP, Galloso I, Feijoo C, Martín CA, Cisneros G (2014) Integration of multisensorial stimuli and multimodal interaction in a hybrid 3DTV system. ACM Trans Multimedia Comput Commun Appl 11(1s):16:1–16:22. doi:10.1145/2617992
69. Beerends JG, De Caluwe FE (1999) The influence of video quality on perceived audio quality and vice versa. J Audio Eng Soc 47:355–362
70. Slater M, Usoh M, Steed A (1995) Taking steps: the influence of a walking technique on presence in virtual reality. ACM Trans Comput Interact 2:201–219. doi:10.1145/210079.210084
71. De Ruyter B, Aarts E (2004) Ambient intelligence: visualizing the future. In: Proc. Work. Conf. Adv. Vis. interfaces – AVI'04. ACM Press, New York, p 203
72. Waltl M, Timmerer C, Hellwagner H (2010) Increasing the user experience of multimedia presentations with sensory effects. In: 11th Int. Work Image Anal. Multimed. Interact. Serv. (WIAMIS). IEEE, Desenzano del Garda, pp 1–4

73. Ademoye OA, Ghinea G (2009) Synchronization of olfaction-enhanced multimedia. IEEE Trans Multimed 11:561–565. doi:10.1109/TMM.2009.2012927
74. Murray N, Qiao Y, Lee B et al (2013) Subjective evaluation of olfactory and visual media synchronization. In: Proc. 4th ACM Multimed. Syst. Conf. ACM, New York, pp 162–171
75. Ghinea G, Ademoye O (2012) The sweet smell of success: enhancing multimedia applications with olfaction. ACM Trans Multimed Comput Commun Appl 8:1–17. doi:10.1145/2071396.2071398
76. Pallavicini F, Cipresso P, Raspelli S et al (2013) Is virtual reality always an effective stressors for exposure treatments? Some insights from a controlled trial. BMC Psychiatry 13:1–10
77. (1990) ITU-T J.100 Recommendation. Tolerance for transmission time differences between vision and sound components of a television signal. International Telecommunication Union (ITU) – Telecommunication sector.
78. Slutsky DA, Recanzone GH (2001) Temporal and spatial dependency of the ventriloquism effect. Neuroreport 12:7–10
79. IJsselsteijn WA, de Ridder H, Vliegen J (2000) Effects of stereoscopic filming parameters and display duration on the subjective assessment of eye strain. In: Proc. SPIE 3957, Stereosc. Displays Virtual Real. Syst. VII. pp 12–22
80. Kooi FL, Toet A (2004) Visual comfort of binocular and 3D displays. Displays 25:99–108. doi:10.1016/j.displa.2004.07.004
81. Meesters L, IJsselsteijn W (2003) Survey of perceptual quality issues in threedimensional television systems. Proc. SPIE
82. Banos RM, Botella C, Alcaniz M et al (2004) Immersion and emotion: their impact on the sense of presence. Cyberpsychology Behav 7:734–741
83. Timmerer C, Waltl M, Rainer B, Hellwagner H (2012) Assessing the quality of sensory experience for multimedia presentations. Signal Process Image Commun 27:909–916
84. ISO/IEC 23005-3 (2013) Information technology – Media context and control – Part 3: Sensory information. p 104
85. Waltl M, Rainer B, Timmerer C, Hellwagner H (2013) An end-to-end tool chain for Sensory Experience based on MPEG-V. Signal Process Image Commun 28:136–150
86. Yoon K (2013) End-to-end framework for 4-D broadcasting based on MPEG-V standard. Signal Process Image Commun 28:127–135
87. Kim J, Lee C-G, Kim Y, Ryu J (2013) Construction of a haptic-enabled broadcasting system based on the MPEG-V standard. Signal Process Image Commun 28:151–161

Chapter 3
3D Video Representation and Coding

**Sérgio M.M. Faria, Carl J. Debono, Paulo Nunes,
and Nuno M.M. Rodrigues**

Abstract The technologies which allow an immersive user experience in 3D environments are rapidly evolving and new services have emerged in various fields of application. Most of these services require the use of 3D video, combined with appropriate display systems. As a consequence, research and development in 3D video continues attracting sustained interest.

While stereoscopic viewing is already widely spread, namely in TV and gaming, new displays and applications, such as FTV (Free viewpoint TV), require the use of a larger number of views. Hence, the multiview video format was considered, which uses N views, corresponding to the images captured by N cameras (either real or virtual), with a controlled spatial arrangement. In order to avoid a linear escalation of the bitrate, associated with the use of multiple views, video-plus-depth formats have been proposed. A small number of texture and depth video sequences are used to synthesize intermediate texture views at a different space position, through a depth-image-based rendering (DIBR) technique. This technology allows the use of advanced stereoscopic display processing and to improve support for high-quality autostereoscopic multiview displays.

In order to provide a true 3D content and fatigue-free 3D visualization, holoscopic imaging has been introduced as an acquisition and display solution. However, efficient coding schemes for this particular type of content are needed to enable proper storage and delivery of the large amount of data involved in these systems, which is also addressed in this chapter.

S.M.M. Faria (✉) • N.M.M. Rodrigues
ESTG, Instituto Politécnico de Leiria/Instituto de Telecomunicações, Leiria, Portugal
e-mail: sergio.faria@co.it.pt; nuno.rodrigues@co.it.pt

C.J. Debono
Department of Communications and Computer Engineering, University of Malta, Msida, Malta
e-mail: c.debono@ieee.org

P. Nunes
ISCTE, Instituto Universitário de Lisboa/Instituto de Telecomunicações, Lisbon, Portugal
e-mail: paulo.nunes@lx.it.pt

© Springer Science+Business Media New York 2015   25
A. Kondoz, T. Dagiuklas (eds.), *Novel 3D Media Technologies*,
DOI 10.1007/978-1-4939-2026-6_3

3.1 Introduction

Telecommunication networks are experiencing large demands for video data, where IP video traffic is expected to be around 79 % of all consumer Internet traffic by 2018 [1]. These requirements will be even more stressed with the event of Immersive multimedia services which need the transmission of huge amounts of data, as multiple camera video content must be transmitted to better render the three-dimensional (3D) scene at the display. The uptake of this technology will further push the amount of IP video traffic passing over the networks. This 3D content has to pass over bandwidth-limited channels and to be stored on space-limited hardware. Some 3D video content is already being transmitted, with current technology being limited to stereoscopic video over satellite channels, stored content on Blu-Ray^{TM} disks, and Internet technologies [2]. Therefore, in order to sustain the massive growth in communications needs imposed by the transmission of 3D video content, adequate video coding techniques are needed. The choice of the video coding scheme that is used for a particular application or service depends on the data format and the hardware that is available at the display. This means that adequate design of video representation methods is needed to facilitate the speed and amount of compression that can be achieved during 3D video coding.

In this chapter we will first describe the different 3D video representation formats starting from stereo and then moving to multiview and holoscopic formats. Emerging formats of multiview video-plus-depth and layered depth video are also discussed. The chapter then focuses on the different techniques used to encode 3D videos. These include stereoscopic, multiview, video-plus-depth, and non-scalable 3D holoscopic video (3DHV) coding. A final conclusion is then drawn together with some discussions on future trends.

3.2 3D Video Representation

This section presents an overview of the existing formats to represent 3D video. All of these formats aim to represent the captured scene from two or more viewpoints. Original systems used a stereoscopic video representation, consisting of two views of the captured scene. When more than two views are used, we refer to the representation as multiview video (MVV). This allows the representations of the captured scene from any viewpoint, or its display in autostereoscopic displays. When the number of used views is very large (hundreds or thousands), we refer to it as holoscopic video.

Alternative representations may use information beyond the captured views. An option that has recently gained a great relevance is the use of depth information associated with a texture view, which is commonly referred to as video-plus-depth.

Recently proposed formats aim to represent more complex 3D models of the objects in the scene, usually by using a large number of views combined with image rendering algorithms. Two examples of these formats are the representation of the scene through the use of a plenoptic image and the use of layered depth video.

3.2.1   From Stereo to Multiview and Holoscopic Video

Providing a more immersive and closer to reality, multimedia experience to home-users has attracted the attention of many researchers for a long time. Stereoscopic 3D systems provide an additional level of immersiveness relatively to 2D, due to the enhanced depth perception obtained from the extra view presented to the user. However, stereoscopic systems do not provide motion parallax, as the user is presented always with the same perspective of the scene, independently of his/her relative position to the display system. This limitation can be partially solved through MVV systems that try to represent multiple perspectives of the same scene, allowing the user to see a different perspective depending on his/her relative position to the display system.

In order to overcome the disadvantages related to human factors of available 3D video technologies, intense research in new types of 3D content representation has continued. Due to recent advances in theory, microlens manufacturing, sensor resolutions and display technologies, 3D holoscopic imaging, also known as integral imaging, light-field imaging, or plenoptic imaging, is becoming a practical prospective 3D technology, which is able to create a more realistic depth illusion than stereoscopic or multiview solutions and, thus, promises to become a popular imaging technology in the future.

This section intends to review in a comprehensive way the basics of 3D content representation, from stereoscopic to multiview and holoscopic representations.

3.2.1.1   Stereoscopic 3D Video

Generically, 3D video stands for any video format that supports the ability to present "slightly" different motion images (views) of the visual scene to each of the viewers' eyes. The binocular disparity induced by the different images reaching each eye is used by the brain to provide enhanced depth perception [3].

This means that 3D video is usually denoted as a general video representation conveying multiple views of the visual scene to the receiver. The particular case of 2-view video formats is commonly referred to as stereoscopic 3D video (S3DV or just S3D) or simply stereoscopic video. Figure 3.1 illustrates a stereo image pair and the corresponding difference image, where the horizontal disparities can be easily observed.

S3DV is, thus, the simplest 3D video format both in terms of acquisition and also in terms of transmission and display. The image separation at the display side can be done in a variety of ways [4], being the active shutters (time multiplexing) and the polarized glasses (spatial multiplexing) the most common in cinema and home

Fig. 3.1 Stereoscopic image pair and corresponding difference image

environments. Nowadays, autostereoscopic systems that don't require any eyewear device are emerging, though not quite disseminate in the consumer market. In this case, image separation is performed optically using parallax barriers or lenticular systems [4, 5].

3.2.1.2 Multiview Video

When more than two views are conveyed to the receiver, the 3D video format is simply referred as MVV. In this case, MVV acquisition is typically performed by an array of synchronized cameras displaced horizontally [6].

Using appropriate multiview displays, motion parallax is also supported, besides stereo parallax, meaning that the user is able to see a different scene perspective when moving the head along the horizontal plane. This is usually achieved through multiview autostereoscopic displays with or without head tracking [4].

3.2.1.3 3D Holoscopic Video

When a dense number of views is captured simultaneously, MVV can be understood as a light-field or plenoptic function sampling [7, 8]. This means that not only spatial information about a scene is captured but also angular information, i.e., the "whole observable" (holoscopic) scene.

The concepts of 3D holoscopic imaging were proposed over a century ago by Lippmann and referred to as integral photography [9]. In 3D holoscopic imaging a regularly spaced array of small spherical micro-lenses, known as a "fly's eye" lens array [10], can be used for both acquisition and display of the 3D holoscopic images to capture and reconstruct the visual scene. At the acquisition side, each micro-lens works as an individual small low resolution camera conveying a particular perspective of the 3D scene.

At the display side, a similar optical setup may be used, e.g., a high resolution LCD panel overlaid with a micro-lens array, to project the recorded 3D holoscopic image and reconstruct the 3D scene in front of the viewer with both stereo and

Fig. 3.2 Different focal planes extracted from a 3D holoscopic image

motion parallax. More details on 3D holoscopic imaging can be found in the Chapter "3D Holoscopic Video Representation and Coding Technology."

Besides a 3D representation of the visual scene, new degrees of freedom in terms of content manipulation are also possible with 3DHV, such as changing the plane of focus or the scene perspective (see Fig. 3.2).

3.2.2 Multiview Video-Plus-Depth

Recent displays enable a more immersive experience by using more than two views. Current autostereoscopic displays use from about ten to thirty views to achieve a 3D effect without requiring the use of glasses. On the other hand, several emerging applications, like free viewpoint video and free viewpoint TV, require the availability of a large number of views in the decoder/display side.

The straight forward approach of transmitting all the views to the decoder has a requirement that grows linearly with the number of used views, both in terms of the number of used cameras and the bandwidth for transmission or storage space [11].

A more efficient representation of the three-dimensional scene can be achieved by using information about the location of the observed objects in relation to the cameras' position. For this purpose, a representation of the relative depth of each of the areas of the image is used. This method is the so-called multiview video-plus-depth format (MVD) [11, 12]. In MVD, a small number of texture views are used, combined with the geometric information of the scene, represented by *depth-maps*. A depth-map contains the distance of each pixel represented in the texture of the captured view relative to the camera.

In a given acquisition system, the range of depths (z_{min}, z_{max}) is mapped into the values of each element of the bi-dimensional matrix which corresponds to the depth-map. For the common case of a representation using 8 bits for each depth sample, this means that the (z_{min}, z_{max}) range is divided into 256 uniform intervals. An inverse notation is common, meaning that a value of 0 in the depth map corresponds to z_{max}, while z_{min} is represented by a sample value of 255. Figure 3.3 represents the texture captured by camera 3 for the Ballet sequence (left) and its associated depth-map (right).

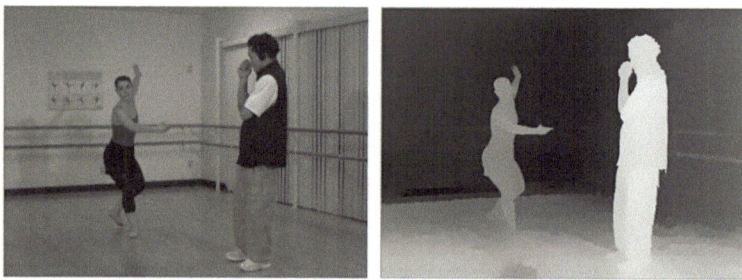

Fig. 3.3 Representation of the texture (*left*) and associated depth-map (*right*) for camera 3 of the Ballet sequence

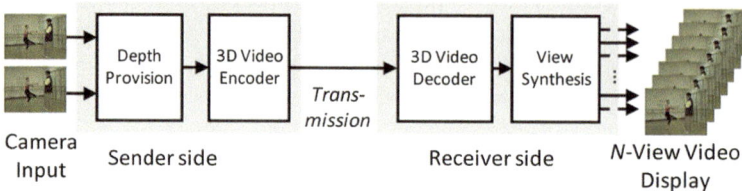

Fig. 3.4 MVD system based on DIBR for view synthesis in the decoder [11]

MVD uses a fixed number of texture views together with the associated depth information. Depth-image-based rendering (DIBR) techniques are then used in the decoder side, to combine the information of the transmitted views and their associated depth-maps, in order to synthesize the intermediate views required by the display. This process is represented in Fig. 3.4. The limited number of encoded views reduces the bit-rate required to transmit the 3DV signal, while the use of DIBR allows the generation of a large number of views in the decoder.

The depth information, z, is available for 3D synthetic sequences, which are based in a geometric model [11]. For natural scenes, the depth value can be determined based on the disparity, d, and the information about the geometry of the acquisition system (namely the baseline distance for the cameras, b, and the focal length, f), by: $z = f \cdot b/d$. The accuracy of the depth value depends on the disparity estimated from the texture views. A large number of disparity estimation methods have been proposed in the literature (e.g., [13–15]). In spite of the improvements in the accuracy of these methods, errors in the determined depth values are still common, especially for totally of partially occluded areas of the 3D scene. These errors limit the applicability of the estimated depth-maps, due to the impact of depth errors in the view synthesis process [11].

In recent years, depth sensors, such as time-of-flight cameras have entered the consumer market. These sensors are currently only capable of acquiring low-resolution depth maps, which are usually enhanced by post-processing methods based on interpolation and denoising filters. Furthermore, these sensors usually present a limited depth range (both for the minimum and maximum values) and, since they are independent from the texture cameras, they have to be placed at slightly different positions.

3.2.3   Layered Depth Video

The layered depth video (LVD) is a 3D video representation that comprises texture images, depth maps, and an additional layer providing occlusion information, which is used to enhance the rendering process in the occluded regions [16, 17]. This is a derivative and an alternative of the MVD representation. The MVD only transmits one or more views (with the associated depth maps) and the non-transmitted side views are generated by projection of the central view, using an image-based rendering process.

The main problem is that every pixel does not necessarily exist in every view, often resulting in the occurrence of holes in the synthesized views. The synthesis exposes the parts of the scene that are occluded in the transmitted view and make them visible in the side views, as can be seen in Figs. 3.5 and 3.6. However, it is possible to overcome this problem by considering transmitting more data

Fig. 3.5 *Top*: Input stereo pair from cam 3 and cam 5

Fig. 3.6 Generation of virtual views for cam 4 from cam 3 or cam 5

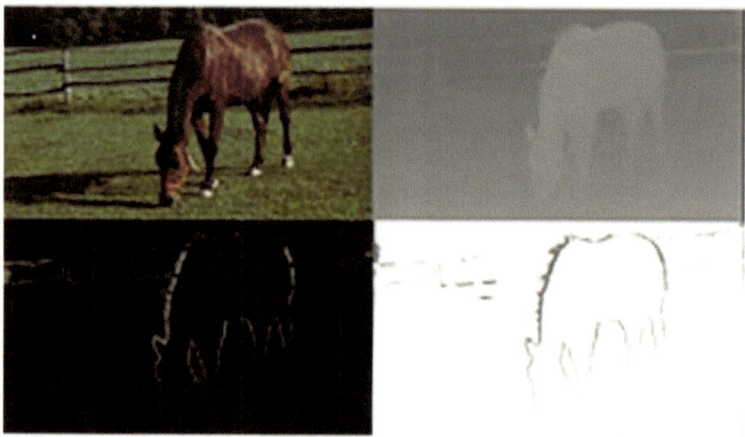

Fig. 3.7 Example of LDV

representation, such as additional texture and depth maps for pixels that are occluded in the transmitted view. This extra data, shown in the bottom pictures of Fig. 3.7, provide the necessary information to fill in the disoccluded regions of the newly generated view.

This representation can be an alternative to MVD that contains a lot of redundant information, as foreground and occluded areas will be visible from multiview points. Thus, the LDV image representation may transmit only one view (with associated depth map) and additional residual data from the occluded areas, thus reducing the required bitrate [18]. This scheme may be extended to include two sets of LVD, which is referred to as depth-enhanced stereo (DES) [19]. This representation format is more suited for stereoscopic glasses-based consumer displays, increases the occlusion information, and allows interpolation or extrapolation of the views.

3.3   3D Video Coding

Despite the potential of the stereoscopic and MVV formats, for a variety of applications, all of these formats require a much larger amount of data than the traditional 2D representation of a scene. The escalation of the requirements associated with storage and transmission of these sequences has motivated the development of many different algorithms to encode 3D video. In this section, we present an overview of the most important coding algorithms for 3D video sequences, using the representation formats described in Section 2.

3.3.1 Stereoscopic Video Coding

The initial approach for 3D video coding and transmission involved the use of a couple of cameras to acquire two offset video sequences separately to the left and right eye of the viewer. This is still the most common way to deliver 3D video content, by means of a simulcast or a frame compatible approach. In this section we will describe briefly how these systems use the state-of-the-art video coding standards, H.264/AVC [20] and H.265/HEVC [21], to implement such solutions.

3.3.1.1 Frame Compatible

Allowing the transmission of 2D-compatible formats over current networks enabled the rapid deployment of 3D video delivery services and applications. Broadcasters embraced this solution in the initial phase of services, maintaining the use of legacy video decoders with 3D-ready displays already popular in the user market.

Therefore, a set of frame compatible formats have been defined, which allow to transmit stereoscopic video without the need to modify the coding algorithms used for single video compression, like H.264/AVC or H.265/HEVC. For a stereo video, it consists of packing pixels from left and right views into a single frame of compressed stream, at the cost of reducing the spatial or the temporal resolution. In the former, the spatial resolution is affected due to the column or row pixel downsampling in each view, while in the latter the views are interleaved as alternating frames or fields, as can be seen in Fig. 3.8.

The left and right views are identified in the compressed stream via specific signaling. 3D frame compatible formats encoded with H.264/AVC use the supplemental enhancement information (SEI) messages, as defined by the subclause D.2.25 in the standard [20, 22]. These messages are used to convey information

Fig. 3.8 Example of frame compatible formats. (**a**) Side-by-Side; (**b**) temporal interleaving

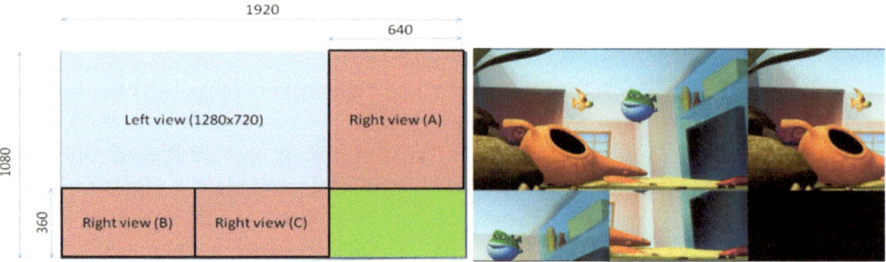

Fig. 3.9 Example of tile compatible format [23]

about how decoders should handle their output according to the used frame-packaging scheme. The various packing formats include: Checkerboard, column-based interleaving, row-based interleaving, top–bottom, temporal interleaving, and 2D. In the last amendment of H.264/AVC the tile format [23] was added, which is able to handle two HD frames (1280 per 720 pixels) to be packed into a full HD frame (1920 per 1080 pixels) using a tiling method, where different regions of one view are tiled with the other view.

As seen in Fig. 3.9, the left view is located at the top left corner of the full HD frame and the right view is divided into three regions, which are placed in specific regions of the resulting full HD frame. While maintaining the backward compatibility with legacy 2D devices, as it requires only a 720p crop to obtain a 2D version of the content in HD resolution, the full resolution of the original frames is maintained in all dimensions, as there are no downsampling operations involved. Also in H.265/HEVC, the spatial and temporal frame packing arrangement is supported by SEI messages.

3.3.1.2 Simulcast with Hybrid Schemes

Despite the higher compression efficiency of the most advanced video coding algorithms, the providers must take into consideration the capabilities of the existing receivers and the transition plan, when introducing new services. For example, there are still many terrestrial broadcasting systems based on MPEG-2. Also several US cable providers rely their systems on mix of MPEG-2 and H.264/AVC. Thus, rather than using the video encoder algorithm for the second view, it is also possible to consider the use of advanced codecs to encode the video required to form the 3D program. This is referred to as a hybrid solution, since two different codecs (e.g., MPEG-2, H.264/AVC and H.265/HEVC) would be used to represent a single 3D program [24].

With the advent of H.265/HEVC it is likely that this high efficiency video coding algorithm may be preferred to encode the second view, independently of the residue resulting from some kind of inter-view or temporal prediction, as depicted in

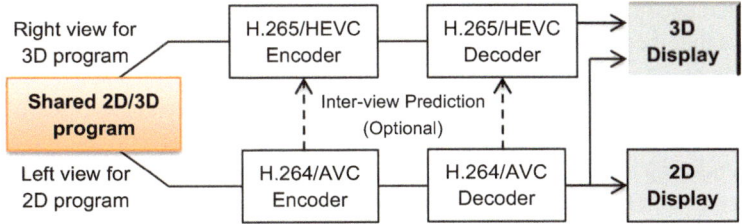

Fig. 3.10 Hybrid architecture based on H.264/AVC

Fig. 3.10. However, while in the former scheme two codecs are required, one for each view, in the latter a new standardization is needed.

3.3.2 Multiview Video Coding

Multiview video coding (MVC) was developed as an extension of the monoview (or monoscopic) H.264/AVC video coding standard. MVC provides a compact representation for multiple views of a video scene without compromising spatial resolution or video quality, relatively to frame-compatible formats. To achieve high compression efficiency, inter-view prediction is also used in addition to existing spatial and temporal predictions. The basic concept of inter-view prediction, introduced in the multiview profile of the MPEG-2 video coding standard [25] and further developed in H.264/AVC, is to exploit the existing redundancy between multiple views of a video scene, since in a multiview scenario the various views of the video scene are typically captured by nearby viewpoints, thus exhibiting substantial inter-view redundancy [26]. With the advent of new extensions of the HEVC standard, multiview video coding has been further developed in a display scalable way.

Nevertheless, MVC involves coding multiple views, increasing linearly the required bitrate with the number of views. This solution is acceptable for stereo or for a system using a reduced number of views. When a large number of views are needed, it becomes impracticable. For such requirements, a different approach was devised based on the synthesis of intermediate views in the decoder, thus transmitting fewer views. As the synthesis requires depth information, a video-plus-depth format was proposed, as explained in the following section.

3.3.3 Video-Plus-Depth Coding

Transmission of 3D content using multiple views demands large bandwidths even when considering a small number of views. However, the display equipment, such as autostereoscopic displays, requires simultaneous decoding of a number of views. It follows that the decoder should be capable of rendering as many views as possible to provide an immersive 3D experience. To satisfy these needs, view synthesis can be employed at the receiver to generate the necessary views by interpolation from the actually received views. A commonly used synthesis algorithm is depth-image-based rendering (DIBR) [27], where the depth maps are used to provide the necessary geometry during the interpolation process. Depth estimation is a high computational task, which cannot be performed at the decoder in real-time applications [28] and, as a result, has to be transmitted together with the texture video. This gives rise to the video-plus-depth coding schemes discussed in this section.

3.3.3.1   Depth Map Coding Techniques

The search for standardization of efficient compression techniques for MVD is being conducted by the Joint Collaborative Team on 3D video coding extension development (JCT-3 V). JCT-3 V is composed by elements from MPEG (ISO/IECJTC1/SC29/WG11 3DV ad hoc group) and ITU-T (SG 16 WP 3). The compression of MVD is in line with most previous video coding standardization efforts and most of the current proposals are being developed in a H.264/AVC or H.265/HEVC framework. Nevertheless, several proposals for depth-map coding (DC) techniques have been made outside of the scope of standardization groups. In this section we will review the most relevant techniques for DC.

Depth-maps are bi-dimensional matrixes of elements that represent the depth associated with each pixel of the associated texture view. For this reason, DC is often regarded as an image compression problem. Nevertheless, efficient DC techniques must take into account the particular features of depth-maps. Depth-maps are commonly composed by large homogeneous areas, associated with the objects at a given depth-plane, surrounded by sharp edges, corresponding to the objects' boundaries. Since depth-maps will be used for DIBR, and not for display, the compression efficiency for DC should be measured based on the distortion of the synthesized intermediate views. DIBR is particularly sensitive to errors in the edges of the compressed depth-maps. These edges correspond to high frequency regions, which are strongly affected by traditional transform-based encoders, and should be taken into account during the development of DC.

The MPEG standardization groups are working on AVC and HEVC-based encoders for depth-maps. For AVC, the encoder is referred to as 3D-AVC [29]. The extension to the HEVC standard supporting DC is known as 3D-HEVC [30, 31]. Both these encoders incorporate specific tools for the independent compression of depth-maps or for the joint compression of texture-plus-depth.

The 3D-HEVC techniques for DC are based on the traditional hybrid video encoding model, which uses predictive coding combined with transforms and entropy encoding [12]. 3D-HEVC uses specific encoding setups (e.g., disabling the de-blocking filter and the sample adaptive offset filter), but new tools for DC have also been developed, including: depth modeling modes, region boundary chain code, and simplified depth coding [23, 32–34]. Depth Lookup Table [33, 35] reduces the bit depth of the each depth-map sample, achieving good compression performances for highly quantized depth-maps, which are commonly used. View Synthesis Optimization (VSO) [33, 34] performs rate-distortion optimization for the compression of the depth-map samples based on the distortion between two intermediate views, synthesized with the original and the compressed depth-maps.

The use of intra prediction is a powerful tool for image compression and has also been adopted for DC. Due to the differences between texture and depth-map images, several proposals have been made for intra prediction modes specifically tailored for DC. These modes can be used to improve the performance of traditional intra prediction framework of H.264/AVC or H.265/HEVC. Intra prediction modes for DC are designed to efficiently represent both the homogeneous areas as well as the edges, which are commonly found in depth-map macroblocks (MB).

3D-HEVC uses the previously mentioned depth modeling modes (DMM) [23, 34], together with the original HEVC intra prediction modes. Furthermore, 3D-HEVC also defines an alternative way to approximate the prediction residues, referred to as constant partition value (CPV). CPV can be used as an alternative to the traditional transform coding. DMM assume that the depth-map blocks are composed by two uniform regions, divided by one or more edges. Explicit information is transmitted to signal the MB structure, as well as the values of the uniform areas. Two types of partitions are defined, namely Contours and Wedgelets. Wedgelets (see Fig. 3.11) assume that a single linear edge divides the two uniform areas, P_1 and P_2. 3D-HEVC defines two wedgelet modes: one explicit and one implicit mode. The former transmits the information about the edge location and values of the uniform regions, while the latter predicts the wedgelet partition based on the texture information. The contours-based prediction modes (see Fig. 3.11) are able to represent more complex arrangements of arbitrarily-shaped regions. The complex region partition is determined based on the texture block in order to avoid the transition of this information. The values of the uniform depth regions, P_1 and P_2, are predicted from previously reconstructed blocks and represented using a Depth Lookup Table (DLT) [33, 36].

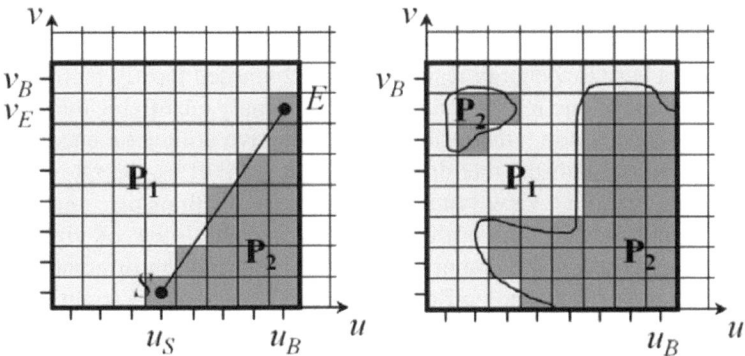

Fig. 3.11 The wedgelet (*left*) and contour (*right*) depth modeling modes of 3D-HEVC [12]

Other authors have proposed DC techniques that include specifically designed intra prediction methods. In [37], a new prediction method is combined with H.264/AVC, which defines two regions for the depth-map macroblocks (MB), which are represented by a losslessly encoded binary or predicted from previously encoded edge MBs. [38, 39] also proposed an additional intra mode for the H.264/AVC encoder, which adaptively segments each depth MB in order to capture the edge structure. A constant [39] or a planar [38] approximation is then used for each segment. In [37] the authors used an approximation of MB segments with a constant surface for HEVC-based DC.

In spite of the advantages of these edge aware intra prediction modes for DC, the use of transform coding for compressing the prediction residue often introduces coding errors in the high frequency components, i.e., the edges. This motivated the proposal of several techniques which exploit new paradigms for DC. Probably the most well know is the Platelet encoder [40]. This algorithm segments the depth MB using a combination of quadtree decomposition with arbitrary segmentation using a linear edge, and approximates each segment by a piecewise-linear function. This efficiently preserves sharp object boundaries and edges, resulting in a high rendering quality.

Graziosi et al. [41, 42] describe the use of an alternative coding paradigm for DC, based on pattern matching, with state-of-the-art results. The proposed method is referred to as Multidimensional Multiscale Parser (MMP) [43]. Each MB is first predicted using a set of prediction modes based on the ones of the H.264/AVC standard. The resulting residue is then encoded by using approximate pattern matching with codewords from an adaptive dictionary. The pattern matching step uses blocks of different sizes (scales), through the use of an appropriate scale transformation, which is applied to the dictionary elements prior to the block matching.

In [36, 44], a DC algorithm that shares some common points with MMP is presented. The Predictive Depth Coding algorithm (PDC) is based on the same flexible partitioning used by MMP [43], combined with an improved predictive framework and a new residue encoding scheme. In [36] the authors demonstrate the advantages of using large MB sizes, combined with a very flexible partitioning scheme. The large MBs are able to efficiently approximate the homogeneous regions of the depth-maps. The highly flexible partitioning scheme of PDC is able to efficiently capture the edge structure of the depth-maps. The result is the generation of a very low energy residue signal, composed mostly by null samples. The few non-null residue regions are encoded by simply quantizing the mean value of the residue and encoding the result with an adaptive arithmetic encoder.

Figure 3.12 represents the rate-distortion results (PSNR vs. the rate required for DC) for one intermediate view synthesized using the depth maps compressed with some of the discussed DC algorithms. The intermediate view of virtual camera 4 was synthesized using the compressed depth-maps and the original textures of views 3 and 5 of the well-known Ballet (left) and Breakdancers (right) sequences. Results for the Platelet [40], MMP [42], and PDC encoder [36] are compared with the state-of-the-art, transform-based H.264/AVC and H.265/HEVC for the intra compression of depth-maps. Both plots show the consistent advantage of PDC algorithm when compared with other state-of-the-art depth map coding algorithms.

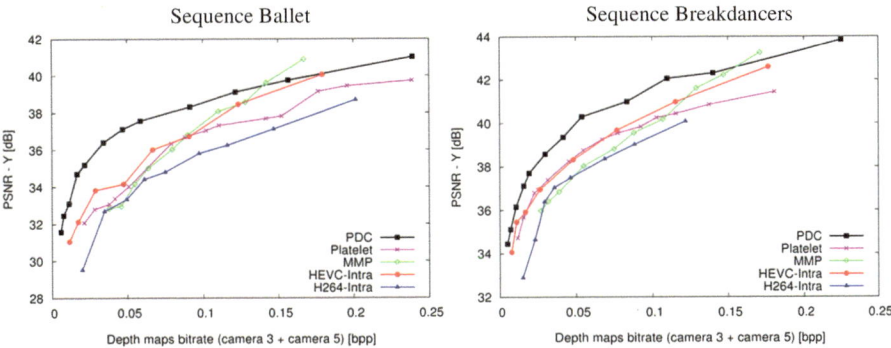

Fig. 3.12 PSNR results of the intermediate view (virtual camera 4) synthesized using the compressed depth maps together and the original texture views (3 and 5)

3.3.3.2   Joint Coding Techniques for Texture and Depth Maps

The multiview video-plus-depth format, which consists of the texture and the depth map MVV data allows for better reconstruct of the 3D video at the receiver. The texture MVVs represents the scene as captured from fixed positions, while the depth MVVs provide the geometric information that allows depth image-based rendering of virtual texture viewpoints at any arbitrary position and the reconstruction of better quality 3D. While allowing for the transmission of a smaller set of views, the additional depth MVVs still need a portion of the channel bandwidth available. Therefore, for 3D video technology to be successful, efficient MVC of both texture and depth MVVs is necessary. In this section we will describe the joint coding techniques used for texture and depth MVVs. We will explore methods to improve encoding speed and coding efficiency in multiview video-plus-depth.

The multiview video coding scheme has been used extensively in literature to compress the video-plus-depth videos. This would generate two MVC compatible streams, one representing the texture MVVs and the other the depth MVVs. The depth MVVs are considered as luminance video signals when coded using MVC and use the same structure used for the texture MVVs [45]. However, these two MVVs streams contain information and similarities that can be shared to support the coding of the streams.

Fast Video-Plus-Depth Coding

Multiview video coding removes temporal and inter-view redundancies by applying motion estimation and disparity estimation techniques, respectively. These estimations are performed by searching for the best vector that can be used to compensate a macroblock from a reference frame. This reference frame can be either a previously coded frame in the same view or a frame in another view. This vector is searched exhaustively for each sub-macroblock in all potential reference frames and is selected by picking the one that minimizes the rate-distortion (RD) cost [32]. The search starts from the center of a fixed search area, forming a predictor vector, and the selected vector is transmitted as a residual vector with reference to this predictor. This method provides coding efficiency but it is very computationally expensive. Some fast estimation techniques, such as [46], have been applied to H.264/AVC to reduce computations while maintaining good performance.

Observing the disparity estimation method, during this search we are looking for the best disparity vector that minimizes the RD cost function. For a sub-macroblock, the disparity vectors presenting the least distortions should lie in proximity of the corresponding areas in the reference views. This occurs because the different cameras are capturing the same scene from different positions. Therefore, the authors in [33] use the depth data and the multiview geometry to identify these locations in the reference views. The projection matrix P and the object's depth are used to find the location of the object in 3D space and this point is then re-projected to the reference views. This is done using the multiview geometry equation:

$$\zeta m = PM$$

where $m = (u, v, 1)^{\mathrm{T}}$ are the homogeneous coordinates of the top-left corner of the sub-macroblock, $M = (x, y, z, 1)^{\mathrm{T}}$ are the homogeneous coordinates of the 3D point, and ζ represents the depth. The estimate is done on the average values of the corresponding depth video, which implies that the depth maps can be of low resolution. This means that the impact on the extra channel capacity due to the depth video is minimized. The technique results in a more accurate disparity vector predictor and as a result the search area can be reduced. This drastically reduces the disparity estimation computations.

The multiview videos also display the same objects movement from different perspectives, which implies that the motion vectors too have a defined relationship. A better estimate of the motion vector can therefore be obtained from the reference view and the disparity predictor found above is used to locate the sub-macroblock and hence its motion vector. The estimated disparity and motion vectors are still transmitted as residual vectors from the respective median vector maintaining compatibility with the standard. The results presented in [33] show that the search area can be reduced from ±32 to ±10 pixel elements, making the encoder around 4.2 times faster than the exhaustive search.

The authors in [47] demonstrate that the epipolar geometry can be used to limit the disparity estimation search area. The epipolar geometry describes the relationship between two multiview images, which dictates that a point $m = (u, v, 1)^{\mathrm{T}}$ must lie on the epipolar line in the reference views:

$$Fm = l$$

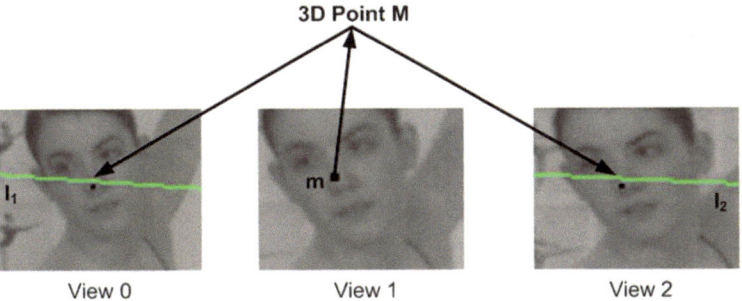

Fig. 3.13 The relation between the current macroblock and the epipolar lines in the reference views

where $\mathbf{F}$ is the (3×3) fundamental matrix and $\mathbf{l} = (a, b, c)^{\mathrm{T}}$ represents the line having the form $ax + by + c = 0$. $\mathbf{F}$ can be found using the camera projection matrices or the methods described in [48]. This relation suggests that the disparity estimation area can be further reduced to a small region around this line. Figure 3.13 shows an example of the current view (view 1) and its references (view 0 and view 2) of the *Ballet* sequence. Results given in [49] show that a speed up of 21.8 times can be achieved if the epipolar line is considered together with the disparity vector discussed above.

The geometry methods were further enhanced in [50] by using an adaptive search area along the epipolar lines based on the largest depth variation within the macroblock being encoded. A speedup of 32 times is reported over the exhaustive full search algorithm. The equation used to determine the search area is given by:

$$\mathrm{SA} = \pm \min\left(3 + \alpha \cdot \Delta_{\mathrm{depth}} \cdot 10\right)$$

where Δ_{depth} is the absolute difference between the statistical maximum and minimum depth value within the sub-macroblock, while α is sequence dependent and depends on the statistical variation of the depth within the sub-macroblock. Using 8×8 pixel elements, this parameter is found as:

$$\alpha = 7/\sigma$$

where σ is the standard deviation of the differences between the statistical maximum and minimum depth values.

The authors in [51] propose an inter-component tool for the HEVC standard to perform joint coding of the quadtrees in texture and depth videos. This technique saves coding time and provides better compression, hence reducing the data that needs to be transmitted. When the depth video is encoded before the texture video, the quadtree structure for the texture is inherited from the already coded depth quadtree. On the other hand, if the texture is encoded first, the depth quadtree inherits the information from the texture.

All the techniques used to provide a faster encoder result in some loss in coding gain. This occurs because of the predictive nature of the used algorithms that are sub-optimal in nature. The losses are well contained and the gain in terms of speed is more beneficial than the small loss in quality, which is typically difficult for the viewer to distinguish.

Improved Coding Efficiency

The 3D video coding extension (3DVC) of the high efficiency video coding that is currently being developed by MPEG takes into consideration the similarities between the texture and depth MVVs during encoding. The base view can be independently decoded with a HEVC codec for compatibility purposes but the other views, including depth MVVs, depend on this stream. Additional coding tools are therefore applied together with an inter-component prediction scheme that uses the already coded components at the same time instant [24]. These modifications include motion parameter inheritance where the depth map coding inherits the partitioning of the blocks and the motion parameters from the texture image. These motion vectors are quantized to full-sample precision compared to the quarter-sample accuracy in the texture video. The decision on whether to inherit the vectors or not is done for each block. The rate distortion optimization takes into consideration the synthesized view, given that the main need for multiview video-plus-depth coding is to generate virtual views for autostereoscopic displays and free viewpoint applications. For a more detailed understanding of the multiview video and depth data coding for 3D HEVC, the reader is referred to [24].

In order to save bandwidth, the depth video can be down-sampled at the encoder and then up-sampled again at the decoder. A joint texture/depth edge-based up-sampling algorithm is proposed in [52] to improve the quality of the edges in the depth video and hence provide a gain in quality of the synthesized views. The joint algorithm exploits the similar geometry and intensity of the depth data and the corresponding texture image. Each sample is further adapted using weights to the characteristics of the depth map to further improve the quality. Results reported in [52] on H.264 show an improved coding efficiency.

Other work was directed to use Scalable Video Coding (SVC) to aid multiview video-plus-depth coding. In [46, 53] the authors present an inter-view prediction scheme which is similar to MVC together with an inter-layer motion prediction scheme adopted from SVC. The latter exploits the correlation that exists between object movements in both texture and depth video streams, thus the base layer is used to carry the texture video while the coarse granularity scalability enhancement layer carries the depth data. An improvement of 10–20 % in coding efficiency is reported in [53] and up to 0.97 dB gain is reported in [54].

3.3.4   Non-scalable 3DHV Coding

In order to transmit 3DHV content over limited-bandwidth networks with an adequate quality, efficient video coding tools are needed that fully exploit the inherent spatial and temporal correlations existing in this type of content.

The planar intensity distribution projected behind the micro-lens array, which represents 3DHV frames, consists of a simple 2D array of micro-images of m × n pixels. This is due to the structure of the micro-lens array that is used for capturing this type of content. As such, this 2D array could be simply encoded by any 2D image or video encoder. However, each micro-lens can be viewed as an individual small low-resolution camera, recording a different perspective of the video scene, at slightly different angles. Due to the small angular disparity between adjacent micro-lenses, a significant cross-correlation exists between neighboring micro-images (see Fig. 3.14). Therefore, this inherent cross-correlation of 3D holoscopic images can be exploited for improving coding efficiency. Additionally, a significant correlation also exists between neighboring pixels within each micro-image.

3.3.4.1   3D-DCT Encoding

Early schemes for 3D holoscopic image and video encoding proposed in the literature were based on the three-dimensional discrete cosine transform (3D-DCT) [55–58]. These schemes take advantage of the existing redundancy within the micro-images (i.e., images formed behind each micro-lens), as well as the redundancy between adjacent micro-images, by applying the 3D DCT to a stack of several micro-images.

Other schemes rely, additionally, on the discrete wavelet transform (DWT) [59, 60]. For example, in [59], 3D holoscopic images are decomposed into

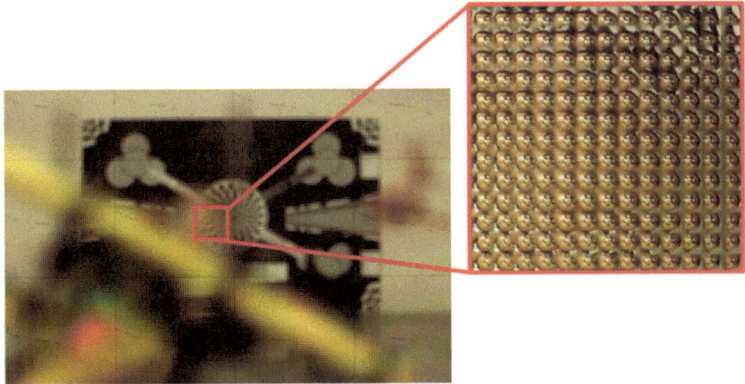

Fig. 3.14 3D holoscopic video frame and image enlargement showing the repetitive structure due to the micro-lens array

sub-images by extracting one pixel with the same position from each micro-lens and each of these sub-images is decomposed using a 2D DWT, resulting in an array of coefficients corresponding to several frequency bands. The lower frequency bands of the sub-images are arranged and transformed using a 3D DCT while the remaining higher bands are simply quantized and the resulting data is entropy encoded.

Encoding schemes for 3DHV are quite scarce in the literature in [61–65], and the authors decompose the 3DHV sequence into several sub-image video sequences and then jointly exploit motion (temporal prediction) and disparity between adjacent sub-images to perform compression. In these schemes, the spatial redundancy is obviously exploited by the disparity estimation part of the scheme, similarly to what is done in MVC; as such, a precise knowledge of the image structure is needed, notably the sub-image dimensions.

3.3.4.2   Self-Similarity Predictive Encoding

A more efficient alternative approach to 3D holoscopic content encoding is to explore the intrinsic redundancy (or self-similarity) of the repetitive structure of this type of images, without needing explicit knowledge of the micro-image structure.

Similarly to motion-compensated prediction encoding, in self-similarity predictive encoding the encoder finds the "best" prediction for each coding unit through a self-similarity estimation process restricting the allowed search area to the already coded and reconstructed areas of the current picture (see Fig. 3.15). The quantized residue and the self-similarity vector representing the relative position between current coding unit and its predictor are then conveyed to the decoder.

The concept of self-similarity predictive encoding was firstly proposed in [66, 67], to improve, respectively, the performance of H.264/AVC for 3DHV coding and to take advantage of the flexible partition patterns used in the emerging HEVC video coding standard.

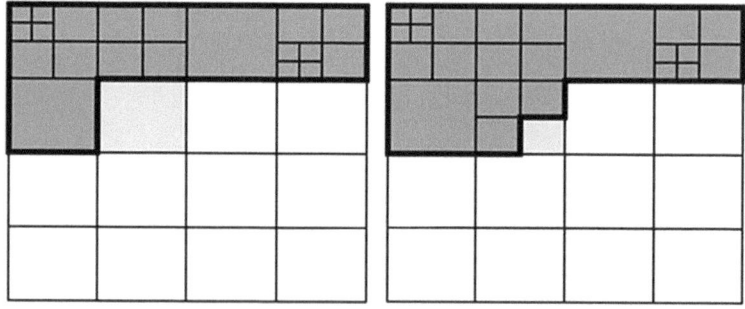

Fig. 3.15 Allowed search areas for estimation of self-similarity vector in two different prediction stages [67]

For more details on Non-scalable and Scalable 3DHV coding, the reader is referred to the Chapter "3D Holoscopic Video Representation and Coding Technology."

Conclusions

This chapter has provided an overview of the current and upcoming formats for 3D video coding, from stereo compatible formats to holoscopic, through multiview video plus depth format. The research on 3D video is progressing at a very rapid pace, driven by academic research as well as standardization efforts and the industrial development of new products for 3D video coding and display.

After presenting the main video representation formats, the main characteristics of the standard coding algorithms have been introduced. This includes the coding tools for stereo images and multiview, as well as for the multiview video-plus-depth. In the latter, some approaches encode the texture and depth information independently, by exploiting the signals' inherent characteristics. Other tools exploit the correlations between the texture video and its associated depth-map, to improve the compression performance. Also distortion models have been described, which evaluate the final distortion on the synthesized views distortion rather than on the depth map distortion.

With the demand for a more immersive experience, and the availability of more advanced displays, which use a very large number of views, the holoscopic format has emerged as a possible solution. However, due to the increased amount of data required, a non-scalable 3DHV coding algorithm is described, for exploiting the cross-correlation that exists between neighboring micro-images. In general, this chapter covers several topics about 3D video representation and coding, providing a set of valuable references, which enables someone interested to get involved in this topic.

Acknowledgements The authors would like to thank the Interactive Visual Media Group of Microsoft Research and National Institute of Information and Communications Technology (NICT), for providing the *Ballet* and *Breakdancers* and *Shark* data set, respectively, for research purposes.

References

1. CISCO (2014) Cisco visual networking index: forecast and methodology, 2013–2018. White paper
2. Vetro A, Tourapis A, Müller K, Chen T (2011) 3D-TV content storage and transmission. IEEE Trans Broadcast 57(2):384–394
3. Zilly F, Kluger J, Kauff P (2011) Production rules for stereo acquisition. Proc IEEE 99(4):590–606

4. Konrad J, Halle M (2007) 3D displays and signal processing. IEEE Signal Process Mag 24 (6):97–111
5. Dodgson NA (2005) Autostereoscopic 3D displays. Computer 38(8):31–36
6. Tanimoto M, Tehrani MP, Fujii T, Yendo T (2011) Free-Viewpoint TV. IEEE Signal Process Mag 28(1):67–76
7. Adelson EH, Bergen JR (1991) The plenoptic function and the elements of early vision. In: Landy M, Movshon JA (eds) Computation models of visual processing. MIT Press, Cambridge, pp 3–20
8. Levoy M, Hanrahan P (1996) Light field rendering. In: Proc. ACM SIGGRAPH. pp 31–42
9. Lippmann G (1908) Epreuves Reversibles Donnant la Sensation du Relief. Journal de Physique Théorique et Appliquée 7(1):821–825
10. Aggoun A et al (2013) Immersive 3D holoscopic video system. IEEE Multimedia 20(1):28–37
11. Muller K, Merkle P, Wiegand T (2011) 3D video representation using depth maps. Proc IEEE 99(4):643–656
12. Muller K, Schwarz H, Marpe D, Bartnik C, Bosse S, Brust H, Hinz T, Lakshman H, Merkle P, Rhee F, Tech G, Winken M, Wiegand T (2013) 3D high-efficiency video coding for multiview video and depth data". IEEE Trans Image Process 22(9):3366–3378
13. Atzpadin N, Kauff P, Schreer O (2004) Stereo analysis by hybrid recursive matching for real-time immersive video conferencing. IEEE Trans Circuits Syst Video Technol 14(3):321–334
14. Lee SB, Ho YS (2010) View-consistent multiview depth estimation for three-dimensional video generation. In: 3DTV-conference: the true vision – capture, transmission and display of 3D video (3DTV-CON). pp 1–4
15. Dongbo Min, Sehoon Yea, Vetro A (2010) Temporally consistent stereo matching using coherence function. In: 3DTV-conference: the true vision - capture, transmission and display of 3D video (3DTV-CON). pp 1–4
16. Shade JW, Gortler SJ, He L-W, Szeliski R (1998) Layered depth images. In: Computer graphics, vol. 32. Annual conference series. pp 231–242 [Online]. Available: http://grail.cs.washington.edu/projects/ldi/
17. Cheng X, Sun L, Yang S (2007) Generation of layered depth images from multiview video. In: IEEE Int. Conf. Image Processing (ICIP), vol. 5, San Antonio, USA
18. Daribo I, Saito H (2011) A novel inpainting-based layered depth video for 3DTV". IEEE Trans Broadcast 57(2):533–541
19. Smolic A, Mueller K, Merkle P, Kauff P, Wiegand T (2009) An overview of available and emerging 3D video formats and depth enhanced stereo as efficient generic solution. In: Picture coding symposium, Chicago, USA, pp 1–4
20. ITU-T and ISO/IEC (2012) Advanced video coding for generic audiovisual services. Rec. ITU-T H.264|ISO/IEC 14496-10
21. ITU-T and ISO/IEC JTC 1/SC 29 (MPEG) (2013) High efficiency video coding. Recommendation ITU-T H.265 and ISO/IEC 23008-2
22. Text of ISO/IEC MPEG2011/N12543 (2012) Additional profiles and SEI messages. San Jose, USA
23. Ballocca G, D'Amato P, Grangetto M, Lucenteforte M (2011) Tile format: a novel frame compatible approach for 3D video broadcasting. In: IEEE International Conference on Multimedia and Expo (ICME), Barcelona, Spain, pp 1–4
24. Müller K, Schwarz H, Marpe D, Bartnik C, Bosse S, Brust H, Hinz T, Lakshman H, Merkle P, Rhee FH, Tech G, Winken M, Wiegand T (2013) 3D high-efficiency video coding for multiview video and depth. IEEE Trans Image Process 22(9):3366–3378
25. ISO/IEC IS 13818-2 AMD3 (1996) MPEG-2 Video Multiview Profile
26. Mueller K, Merkle P, Smolic A, Wiegand T (2006) Multiview coding using AVC. MPEG Doc. M12945, Bangkok, Thailand
27. Oh KJ, Yea S, Vetro A, Ho YS (2010) Virtual view synthesis method and self-evaluation metrics for free viewpoint television and 3D video. Int J Imaging Syst Technol 20(4):378–390

28. Fukushima N, Yendo T, Fujii T, Tanimoto M (2007) Free viewpoint image generation using multi-pass dynamic programming. In: SPIE stereoscopic displays and virtual reality systems XIV, vol. 6490, pp 460–470
29. Hannuksela MM, Chen Y, Suzuki T, Ohm J-R, Sullivan G (2013) 3DAVC draft text 8, JCT-3V document JCT3V-F1002
30. Tech G, Wegner K, Chen Y, Yea S (2013) 3D-HEVC Draft Text 2, JCT-3V document JCT3V-F1001
31. Schwarz H, Bartnik C, Bosse S, Brust H, Hinz T, Lakshman H, Merkle P, Muller K, Rhee H, Tech G, Winken M, Marpe D, Wiegand T (2012) Extension of high efficiency video coding (HEVC) for multiview video and depth data. In: 19th IEEE international conference on image processing. pp 205–208
32. Merkle P, Smolic A, Müller K, Weigand T (2007) Efficient prediction structures for multiview video coding. IEEE Trans Circuits Syst Video Technol 17(11):1461–1473
33. Micallef BW, Debono CJ, Farrugia RA (2011) Exploiting depth information for fast motion and disparity estimation in multiview video coding. In: 3DTV conference
34. Muller K, Merkle P, Tech G, Wiegand T (2012) 3D video coding with depth modeling modes and view synthesis optimization. In: Signal Information Processing Association Annual Summit and Conference (APSIPA ASC) 2012 Asia-Pacific, pp 1–4
35. Jager F (2012) Simplified depth map intra coding with an optional depth lookup table. In: International Conference on 3D Imaging (IC3D). pp 1–4
36. Lucas LFR, Rodrigues NMM, Pagliari CL, da Silva EAB, de Faria SMM (2013) Predictive depth map coding for efficient virtual view synthesis. In: IEEE International Conference on Image Processing (ICIP'13), Melbourne, Australia
37. Zamarin M, Salmistraro M, Forchhammer S, Ortega A (2013) Edge-preserving intra depth coding based on context-coding and H.264/AVC. In: IEEE International Conference on Multimedia and Expo (ICME). pp 1–6
38. Oh BT, Wey HC, Park D-S (2012) Plane segmentation based intra prediction for depth map coding. In: Picture Coding Symposium (PCS2012). pp 41–44
39. Shen G, Kim WS, Ortega A, Lee J, Wey H (2010) Edge-aware intra prediction for depth-map coding. In: IEEE International Conference on Image Processing (ICIP2010). pp 3393–3396
40. Merkle P, Morvan Y, Smolic A, Farin D, Mueller K, de With PHN, Wiegand T (2009) The effects of multiview depth video compression on multiview rendering". Signal Process: Image Commun 24(1–2):73–88
41. Graziosi DB, Rodrigues NMM, Pagliari C, da Silva EAB, de Faria SMM, de Carvalho MB (2010) Compressing depth maps using multiscale recurrent pattern image coding. IET Electron Lett 46(5):340–341
42. Graziosi DBG, Rodrigues NMM, Pagliari CLP, Silva E, Faria SMM, Perez MMP, Carvalho M (2010) Multiscale recurrent pattern matching approach for depth map coding. In: Picture coding symposium, Nagoya, Japan
43. Francisco NC, Rodrigues NMM, da Silva EAB, de Carvalho MB, de Faria SMM, Silva VMM (2010) Scanned compound document encoding using multiscale recurrent patterns. IEEE Trans Image Process 9(10):2712–2724
44. Lucas LFR, Rodrigues NMM, Pagliari CL, Silva EAB, Faria SMM (2012) Efficient depth map coding using linear residue approximation and a flexible prediction framework. In: IEEE International Conference on Image Processing (ICIP 2012), Orlando, EUA
45. Merkle P, Smolic A, Müller K, Weigand T (2007) Efficient compression of multiview depth data based on MVC. In: 3DTV conference
46. Zhu S, Ma K-K (2009) A new diamond search algorithm for fast block-matching motion estimation. IEEE Trans Image Process 9(2):387–392
47. Lu J, Cai H, Lou J-G, Li J (2007) An epipolar geometry-based fast disparity estimation algorithm for multiview image and video coding. IEEE Trans Circuits Syst Video Technol 17(6):737–750

48. Hartley R, Zisserman A (2003) Multiple view geometry in computer vision. Cambridge University Press, Cambridge, pp 279–309

49. Micallef BW, Debono CJ, Farrugia RA (2011) Fast disparity estimation for multiview plus depth video coding. In: IEEE visual communications and image processing conference

50. Micallef BW, Debono CJ, Farrugia RA (2013) Low complexity disparity estimation for immersive 3D video transmission. In: IEEE international conference 2013 – workshop on immersive and interactive multimedia communications over the future internet. pp. 622–626

51. Mora EG, Jung J, Cagnazzo M, Pesquet-Popescu B (2014) Initialization, limitation and predictive coding of the depth and texture quadtree in 3D-HEVC. Trans Circuits Syst Video Technol 24(9):1554–1565

52. Deng H, Yu Li, Qui Jinbo, Zhang J (2012) A joint texture/depth edge-directed up-sampling algorithm for depth map coding. In: IEEE international conference on multimedia and expo

53. Zhang J, Hannuksela MM, Li H (2010) Joint multiview video plus depth coding. In: 2010 I.E. 17th international conference on image processing

54. Tao S, Chen Y, Hannuksela MM, Wang Y-K, Gabbouj M, Li H (2009) Joint texture and depth map video coding based on the scalable extension of H.264/AVC. In: IEEE international symposium on circuits and systems. pp 2353–2356

55. Zaharia R, Aggoun A, McCormick M (2002) Adaptive 3D-DCT compression algorithm for continuous parallax 3D integral imaging. Signal Process: Image Commun 17(3):231–242

56. Aggoun A (2006) A 3D DCT compression algorithm for omnidirectional integral images. In: IEEE International Conference on Acoustics, Speech and Signal Processing (ICASSP 2006), Toulouse, France, vol. 2. pp 517–520

57. Zaharia R, Aggoun A, McCormick M (2001) Compression of full parallax colour integral 3D TV image data based on subsampling of chrominance components. In: Proc. of the IEEE Data Compression Conference (DCC 2001), Snowbird, USA, pp 27–29

58. Forman MC, Aggoun A (1997) Quantisation strategies for 3D-DCT based compression of full parallax 3D images. In: IEE International Conference on Image Processing Applications (IPA 1997), Dublin, Ireland, pp 32–35

59. Aggoun A, Mazri M (2008) Wavelet-based compression algorithm for still omnidirectional 3D integral images. Signal, Image Video Process 2(2):141–153

60. Elharar E, Stern A, Hadar O, Javidi B (2007) A hybrid compression method for integral images using discrete wavelet transform and discrete cosine transform. J Display Technol 3(3):321–325

61. Olsson R, Sjostrom M, Xu Y (2006) A combined pre-processing and H.264-compression scheme for 3D integral images. In: IEEE International Conference on Image Processing (ICIP 2006), Atlanta, USA, pp 513–516

62. Adedoyin S, Fernando WAC, Aggoun A (2007) A joint motion and disparity motion estimation technique for 3D integral video compression using evolutionary strategy. IEEE Trans Consum Electr 53(2):732–739

63. Adedoyin S, Fernando WAC, Aggoun A (2007) Motion and disparity estimation with self adapted evolutionary strategy in 3D video coding. IEEE Trans Consum Electr 53(4):1768–1775

64. Conti C, Nunes P, Soares LD (submitted) 3D holoscopic video coding. IEEE Trans Circuits Syst Video Technol

65. Dick J, Almeida H, Soares LD, Nunes P (2011) 3D holoscopic video coding using MVC. In: IEEE International Conference on Computer as a Tool (EUROCON 2011), pp 1–4

66. Conti C, Lino J, Nunes P, Ducla Soares L, Lobato Correia P (2011) Spatial prediction based on self-similarity compensation for 3D holoscopic image and video coding. In: 18th IEEE International Conference on Image Processing (ICIP), pp 961–964

67. Conti C, Nunes P, Soares LD (2012) New HEVC prediction modes for 3D holoscopic video coding. In: 19th IEEE International Conference on Image Processing (ICIP), Orlando, USA, pp 1325–1328

Chapter 4
Full Parallax 3D Video Content Compression

Antoine Dricot, Joel Jung, Marco Cagnazzo, Béatrice Pesquet,
and Frederic Dufaux

Abstract Motion parallax is a key cue in the perception of the depth that current
3D stereoscopy and auto-stereoscopy technologies are not able to reproduce.
Integral imaging and Super Multi-View video (SMV) are 3D technologies that
allow creating a light-field representation of a scene with a smooth full motion
parallax (i.e., in horizontal and vertical directions). However the large amount of
data required is challenging and implies a need for new efficient coding technologies. This chapter first describes integral imaging and SMV content, acquisition,
and display. Then it provides an overview of state-of-the-art methods for full
parallax 3D content compression. Finally, several coding schemes are compared
and a coding structure that exploits inter-view correlations in both horizontal and
vertical directions is proposed. The new structure provides a rate reduction (for the
same quality) up to 29.1 % when compared to a basic anchor structure. Neighboring
Block Disparity Vector (NBDV) and Inter-View Motion Prediction (IVMP) coding
tools are further improved to efficiently exploit coding structures in two dimensions, with rate reduction up to 4.2 % with respect to the reference 3D-HEVC
encoder.

A. Dricot
Orange Labs, Equinoxe 4 avenue du 8 Mai 1945, 78280 Guyancourt, France

Institut Mines-Télécom; Télécom ParisTech; CNRS LTCI,
46 rue Barrault, 75634 Paris, Cedex 13, France

J. Jung
Orange Labs, Equinoxe 4 avenue du 8 Mai 1945, 78280 Guyancourt, France

M. Cagnazzo (⊠) • B. Pesquet • F. Dufaux
Institut Mines-Télécom; Télécom ParisTech; CNRS LTCI,
46 rue Barrault, 75634 Paris, Cedex 13, France
e-mail: cagnazzo@telecom-paristech.fr

© Springer Science+Business Media New York 2015
A. Kondoz, T. Dagiuklas (eds.), *Novel 3D Media Technologies*,
DOI 10.1007/978-1-4939-2026-6_4

4.1 Introduction

The improvement of 3D video technologies tends to provide enhanced immersive experiences to the end users [1]. However, currently available technologies are still limited in several ways. In stereoscopic 3D, the use of glasses introduces a lack of comfort, which is combined with unnatural perception stimuli (e.g., when the eyes converge on an object in front of or behind the screen but accommodate on the screen) that are annoying for the viewer and can even cause eyestrain and headaches. The small number of views displayed with existing glasses-free auto-stereoscopic systems induces artifacts. The viewing zone is also restricted by a limited number of sweet spots and the system cannot provide a smooth motion parallax (i.e., the visualization is not continuous when moving in front of the display).

A study on Super Multi-View video (SMV) and integral imaging has been initiated during the October 2013 MPEG FTV meeting [2]. In these technologies, tens or hundreds of views are used to obtain a so-called *light-field* representation of a scene. A light-field represents the light rays in a 3D scene, and thus is a function of two angles (ray direction) and three spatial coordinates. This five-dimensional function is called plenoptic function [3, 4]. Some of the current 3D technologies artifacts can be eliminated using the light-field representation, as, for example, the vergence-accommodation conflict. As a consequence, several companies are already working on light-field display systems [5]. Light-field displays are glasses-free systems that allow a more realistic visualization. They can provide smooth motion parallax, which is a key element for the perception of depth, in horizontal and potentially vertical directions. Immersive telepresence is a typical target use case, as well as the live 3D broadcast of sport events, like 2020 Olympics in Japan, that could be shot by SMV camera arrays and projected on large SMV display systems at several public viewing facilities in major cities around the world. A large amount of data is needed to create a light-field representation, therefore new efficient coding technologies are required to handle the increasing number of input views and exploit the characteristics of their representations [6].

This chapter is organized as follows. In Sect. 4.2, we first describe the capture and display systems associated with the two main technologies that can provide full parallax 3D video content: SMV and integral imaging. Secondly we give an overview of view extraction methods that allow to convert an integral image to a full parallax multi-view representation. In Sect. 4.3, state-of-the-art methods to encode integral imaging content are presented, followed by standard encoder enhancements needed for full parallax SMV content. Finally, we show a new and efficient inter-view prediction scheme to exploit horizontal and vertical dimensions at the coding structure level, and improvements of inter-view coding tools to exploit the two dimensional structure also at the coding unit level. Conclusions are drawn in section "Conclusion."

4.2   Full Parallax 3D Content

A light-field representation of a scene can be obtained by using several images captured from different points/angles of view. We consider here two technologies that can provide full parallax content: Super Multi-view Video (SMV) and integral imaging, and both are associated with a specific acquisition and display technology and with a specific representation.

4.2.1   Acquisition Technology and Resulting Content

4.2.1.1   Camera Arrays

Super multi-view content can be captured using a camera rig as illustrated in Fig. 4.1. The cameras can be aligned horizontally (providing horizontal motion parallax only) or in horizontal and vertical dimensions in the case of full motion parallax content. The camera system can be arranged in linear, arc, or even sparse alignment. Each camera captures the scene from a different point of view and the resulting content consists of several viewpoint images with horizontal and vertical disparities.

4.2.1.2   Integral Imaging

Integral imaging (or holoscopy) is a technology based on plenoptic photography [7]. Integral imaging acquisition uses a lenticular array set in front of a camera device as illustrated in Fig. 4.2. This lenticular array is composed of a large number of micro-lenses, which can have a round, square or hexagonal shape, and can be aligned in rectangular grid or in quincunx. The resulting captured holoscopic image consists of an array of Micro-Images (MIs, also called elemental images) as illustrated in Fig. 4.3. Each micro-lens produces one MI, and each MI contains

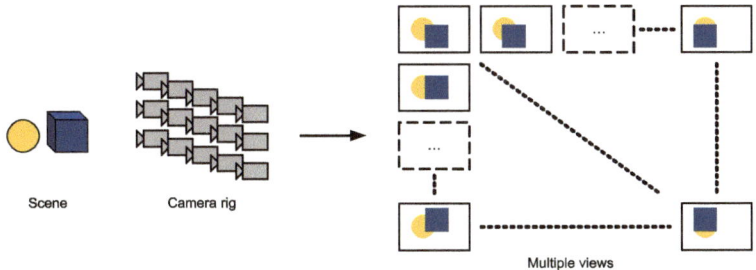

Fig. 4.1 Super Multi-View acquisition using a camera rig

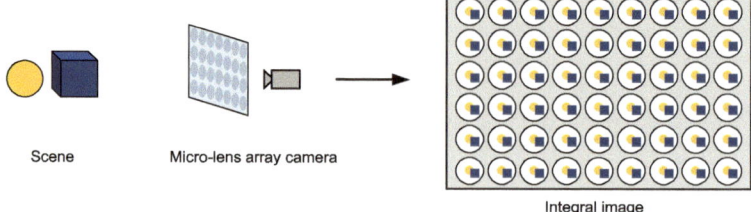

Fig. 4.2 Integral imaging acquisition using a micro-lens array based camera

Fig. 4.3 Close-up on the Seagull plenoptic image [11]

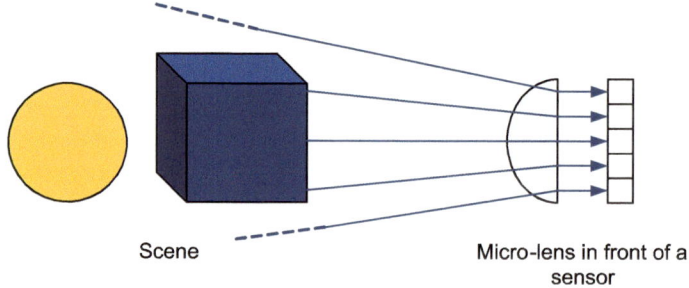

Fig. 4.4 Sampling of angular information by the micro-lens array

the light information coming from several angles of view, as illustrated in Fig. 4.4. Each pixel inside one MI corresponds to a different angle of view of the scene.

In [8], Georgiev and Lumstaine describe the focused plenoptic camera. Traditional plenoptic cameras focus the main lens on the micro-lenses and focus the micro-lenses at infinity. In the focused plenoptic camera the main lens is focused well in front of the lenticular array, which is in turn focused on the image formed inside the camera, so that each micro-lens acts as a relay system of the main lens.

This configuration allows a tradeoff between spatial and angular information inside each MI (see Sect. 4.2.4). Examples of plenoptic handheld cameras exist on the consumer market like Lytro's and Raytrix's products [9, 10].

4.2.2 Display

Several light-field display systems currently exist. There are many experimental or research projects and also commercial products. An overview of these systems has been presented during the October 2013 MPEG FTV meeting [6]. In the following, we classify them into two sections according to the content they take as input: super multi-view content or integral images.

4.2.2.1 Super Multi-View Displays

SMV display systems take tens to hundreds of views as input. Several display systems are based on a front or rear projection as illustrated in Fig. 4.5. For example, this is the case for the REI (Ray Emergent Imaging) system [12] developed by NICT (which uses 170 projection units) or for the Holovizio C80 [13] developed by Holografika (which uses 80 projections units). Each projection unit projects from a different angle onto the screen. The screen surface has anisotropic properties (which can be obtained optically or with a holographic diffuser for example), so that the light rays can only be seen from one direction which depends on the projection direction.

Another kind of system is presented in [6] as "all-around." These display systems are presented in a table-top configuration, hence they offer a viewing angle of 360 degrees to the user who can walk/turn around the device. Some systems use a rotating mirror and/or a rotating holographic diffuser (e.g., the light-field 3D display developed by USC [14], or Holo Table developed by Holy

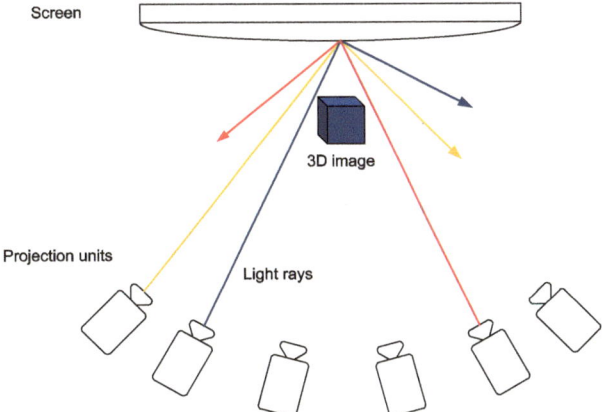

Fig. 4.5 Super Multi-View display system based on front projection

Fig. 4.6 Micro-lens array
based display

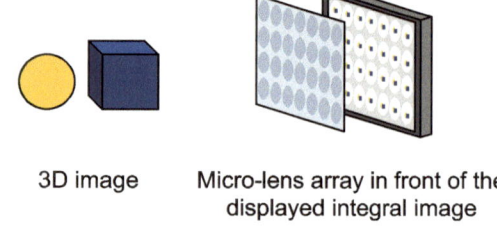

3D image   Micro-lens array in front of the
                    displayed integral image

Mine [15]) while others are based on parallax barrier (SeeLinder developed by
Nagoya University [16]).

4.2.2.2 Holoscopic Displays

Holoscopic display systems use a lenticular array (like the holoscopic capture
systems) setup in front of a screen (see Fig. 4.6). The micro-lenses of the lenticular
array allow the user to see only a particular part of each MI corresponding to the
angle of view (allowing motion parallax). An example of holoscopic display by
NHK is described in [6]. In their standard configuration, holoscopic systems
provide a pseudoscopic image (i.e., with reversed depth) that needs to be converted
to an orthoscopic image. Many improvements are proposed in the literature to solve
the current limitations of holoscopic display systems like the limited depth of field,
limited range of viewing angles, or the conversion from pseudoscopic to orthoscopic image [17, 18]. Among all the display systems cited in [6], only integral
imaging systems have directive properties in the vertical direction hence are able to
display content with full parallax.

4.2.3   Tradeoffs Between Integral Imaging
and Super Multi-View

Integral imaging and SMV technologies both provide light-field representations
because they both sample a subset of the light-field of a scene by capturing images
of that scene from several angles of view. However, this sampling is done in two
different ways, implying the following tradeoffs. Using a camera rig allows to
obtain a wider baseline (e.g., several meters) than using a holoscopic camera for
which the baseline is limited by the size of the micro-lens array. With holoscopic
cameras, the resolution of the viewpoint images is limited because the same sensor
is shared between all the captured views, while with a camera rig the full resolution
of each camera is used for each view. Finally, holoscopic cameras allow a denser
sampling of the light-field, because with a camera rig the distance between each
view is limited by the size of the cameras.

4.2.4   View Extraction

The connection between integral imaging and Super Multi-View can be drawn by the mean of view extraction. Several methods can be used to extract a viewpoint from a holoscopic image. A detailed overview of these methods is given in [19].

4.2.4.1   Basic Extraction

The most basic method to extract a viewpoint image from an integral image is to take only one pixel from each MI, as illustrated in Fig. 4.7. The angle of view can be selected by varying the relative position of this pixel within the MI. The number of angles of view available is equal to the resolution of the MI, and the resolution of the extracted viewpoint images is equal to the total number of MIs. Hence there is a tradeoff between the number and the resolution of the extracted views. This is the main limitation of this method which generally provides views with a small resolution and with pixelation artifacts.

4.2.4.2   View Selective Blending

The goal of this method is to extract intermediary points of view from neighbor views extracted with the previous method. First these views are stacked onto each other, and then they are blended by averaging all the superimposed pixels as illustrated in Fig. 4.8. The resulting extracted image is the viewpoint centered between the stacked views. A drift can be applied to each view in the stack

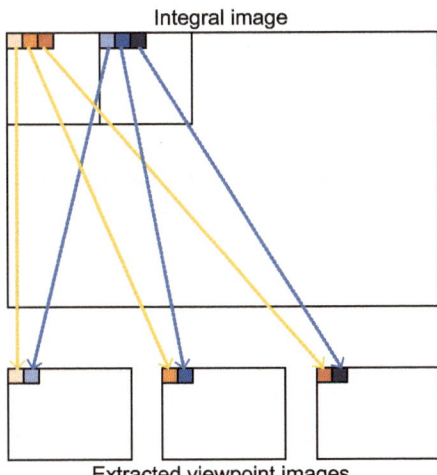

Integral image

Extracted viewpoint images

Fig. 4.7 Basic view extraction method

Fig. 4.8 View selective
blending extraction

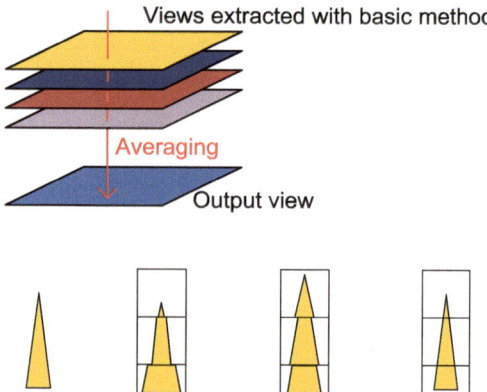

Fig. 4.9 Patch size in view
extraction

(by increasing its distance from the center) before the blending. The value of this drift controls the depth plane on which the resulting extracted view will be focused.

4.2.4.3 Single-Sized Patch

This method proposed by Georgiev is based on the characteristics of the focused plenoptic camera [8] (see Sect. 4.2.1.2) for which there are both angular and spatial information within one MI. It uses the same steps as the basic extraction illustrated in Fig. 4.7, but instead of extracting one pixel per MI, a patch (corresponding to a square zone of pixels) is extracted. The angle of view chosen depends on the relative position of the center of patches in the MIs. The size of the patch defines the depth plane in the scene on which the extracted view will be focused: the larger the patch is, the closer the focus plane is. The objects that are further or closer will present the following artifacts, illustrated in Fig. 4.9. If the patch is too large (i.e., the object is too far), then redundant parts of the object will be represented in several adjacent patches. If the patch is not large enough (i.e., the object is too close), then parts of the object will not be represented (pixelation).

4.2.4.4 Single-Sized Patch Blending

As for the previous method, one patch per MI is extracted. However, the pixels outside the patch's borders are kept to be blended. The blending is done here with a weighted averaging. The weight of pixels inside the MI can be attributed using a Gaussian distribution centered on the patch, so that the pixels that are further from the center have a lighter weight as illustrated in Fig. 4.10. The blending allows

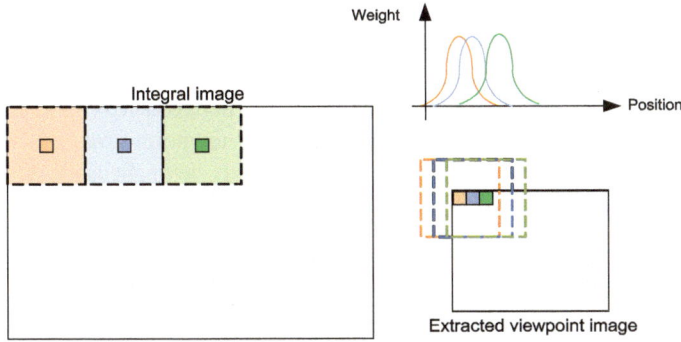

Fig. 4.10 Blending in view extraction

Fig. 4.11 Block matching
algorithm to estimate
disparity between MIs

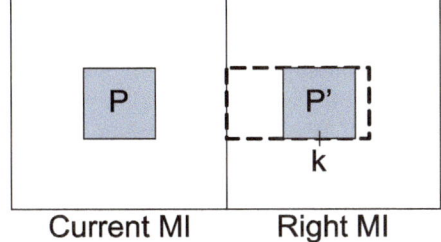

reducing block artifacts by smoothing the transitions between adjacent patches. It
also blurs the parts of the picture that are out of the depth plane that is in focus
(closer or further), which eliminates the mismatch artifacts (described in the
previous section) and provides the same effect as in a classic 2D photograph with
limited depth of field.

4.2.4.5 Disparity Map

A disparity estimation method is proposed in [8] in order to obtain the depth of the
objects inside each MI. It is based on a block matching algorithm (illustrated in
Fig. 4.11) with the following steps. A square patch P is first selected in the center of
the current MI with coordinates (Px, Py), and a second patch P' is selected in a
neighbor MI (e.g., right and/or bottom) with coordinates $P'x = Px + S + k$, with
S the size of the MI and k the disparity between the two MIs. The similitude
between P and P' is computed (e.g., using normalized cross correlation) for values
of k from 1 up to the maximum disparity value possible. The value of k providing
the maximum similarities between P and P' corresponds to the disparity value of the
current MI. This value in number of pixels corresponds to the adequate patch size to
be used for the view extraction. The resulting extracted viewpoint image is full-
focused, as each patch size is adapted to the depth of the objects in the MI.

4.2.4.6 Depth Blending

This technique is a combination of the two previous methods, hence includes varying patch sizes (estimated from the disparity between MIs) and blending. It aims to produce artifacts-free and full-focused viewpoint images. This is the method that provides the most accurate extracted views because it eliminates most of the artifacts reported with less advanced methods.

4.3   Full Parallax Content Compression

4.3.1   State of the Art

4.3.1.1   Integral Imaging Compression

Several methods related to integral imaging compression have been presented in the scientific literature. In the following we group them in five categories: DCT-based methods, Wavelet-based methods, methods based on the processing of micro-images or viewpoint images, methods based on a multi-view approach, and finally the self-similarity (SS) approach.

DCT-Based Methods The most natural approach consists in applying the Discrete Cosine Transform (DCT) to the micro-images, followed by quantization and lossless coding; possibly, a differential coding between MI can be used [20]. The differential coding can also be used for video sequences in order to remove the temporal correlations [21, 22]. Inter-MI correlation can be removed using the 3D-DCT on stacked MIs. Several scanning orders are tested in order to create the MIs 3D structure. An optimization of the quantization step (for 3D-DCT-based compression algorithms) is proposed in [23]. This optimization is done by generating a matrix of quantization coefficients which depends on the content on the image.

Wavelet-Based Methods These methods use a scheme based on a Discrete Wavelet Transform (DWT) applied to the viewpoint images. In [24], a hybrid four-dimensions transform based on DWT and DCT is described (4D hybrid DWT-DCT coding scheme). The 2D DWT is applied to the MIs, followed by a 2D DCT applied to the resulting blocks of coefficients. In [25], the integral image is decomposed in viewpoint images. A 2D transform is performed by using 1D transforms on the lines and rows of the viewpoint images, resulting in four frequency sub-bands. The lower band is a coarse approximation of the original viewpoint image. The 2D transform is applied recursively to increase the level of decomposition at a coarser scale. The sub-bands are then grouped in $8 \times 8 \times 8$ elements volumes and processed by a 3D-DCT. As in the previous methods, the coefficients are then quantized and arithmetically coded. In [26], the transform is combined with a Principal Component Analysis (PCA, also called Karhunen–Loeve Transform or Hotelling Transform). DWT is applied to viewpoint images, and then PCA is applied to the resulting coefficients. Several kinds of DWT filters are proposed (e.g., Dauchechies wavelets). In [27], the SPIHT (Set Partitioning in Hierarchical Trees) method allows to display/transmit progressively the integral image as a quality scalable bitstream. Two algorithms (2D and 3D) are proposed. The first one is a 2D-DWT applied to the integral image and followed by the 2D-SPIHT. The second is based on the creation of a volume of viewpoint images on which a 3D-DWT is applied and followed by 3D-SPIHT.

Elemental Images-Based Methods Another approach consists in encoding the viewpoint images or the MIs of a still integral image as if they were a video sequence (called pseudo video sequence) and then exploiting the temporal compression tools of traditional video coders [28, 29]. This approach shares some ideas with the next two: multi-view and self-similarity. The method proposed in [30] exploits the inter-MIs redundancies (using the optical characteristic that MIs have overlapping zones). In [31], a KLT is applied to the viewpoint images.

Multi-view We consider here the integral image as a group of viewpoint images that is encoded as a multi-view sequence (using inter-view prediction). In [32] and [33], the viewpoint images are encoded using MVC encoder [34]. The exploitation of temporal correlation and inter-view correlations induces an increase in complexity. The evolutionary strategy (ES) proposed in [35] is based on the evolution theory and allows an optimization of the coding scheme. In [36], ES is also applied and combined to a half-pixel precision for the motion/disparity prediction and compensation.

Self-Similarity The method described in this part exploits the non-local spatial correlation between MIs. The algorithm is mainly the same as for the inter prediction modes (of H.264/AVC [37] and HEVC [38]) but within one frame. A block matching algorithm is used to find a block similar to the current block in the causal zone in the current frame (which corresponds to the blocks that have already been coded and reconstructed). This similar block is then used as a predictor in the same manner as for a temporal prediction and compensation. In [39] and [40], the implementation in H.264/AVC of the INTRA_SS (for INTRA Self Similarity) modes is described. These publications show the BD-rate gain brought by the SS mode and also the interest of a precise partitioning in macro-blocks. In [41], the SS mode is implemented in HEVC and the interest of the CU partitioning is shown. Conti et al. [42] shows the BD-rate gain brought by this method for video sequences (i.e., with a competition with inter-prediction modes). In [43] a scalable coding scheme is described as follows: the layer 0 corresponds to an extracted view, the layer 1 corresponds to a set of extracted views, and the layer 2 is the integral image. Layers 0 and 1 are encoded, respectively, with reference HEVC and MV-HEVC encoders. For layer 2 the self-similarity method is used. An inter-layer prediction is also proposed, in which a sparse integral image is reconstructed from the set of views of the layer 1 and is inserted in the reference picture buffer to encode layer 2.

The wavelet-based and 3D-DCT-based methods are conceptually far from the standard encoders schemes (H.264/AVC and HEVC) and more adequate for still image coding, whereas the self-similarity and multi-view methods are more easily included in the structure of these reference video encoders.

4.3.1.2 Multi-View Video Coding Standards

Both H.264/AVC and HEVC standards have multi-view extensions [34]: respectively MVC and MV-HEVC, which provide additional high level syntax that enables the inter-view prediction. 3D-HEVC is an extension for the multi-view plus depth (MVD) format, which provides tools related to depth maps, new tools at the Coding Unit level (CUs in HEVC replace H.264/AVC macroblocks) for side views and new inter-component dependencies. These standard multi-view encoders are designed to handle horizontal parallax content with a small number of views.

The Neighboring Block Disparity Vector (NBDV) [44] and the Inter-View Motion Prediction (IVMP) [45] are specific 3D-HEVC coding tools. In the HTM7.0 reference software of the standard (used for experiments in the following) they are implemented as follows. NBDV searches through already coded temporal and spatial neighboring CUs for a disparity vector (DV) and derives it for the current CU (by simple copy). Motion vectors and temporal references (motion parameters) of the CU pointed by this DV in the reference view are used by IVMP to create the Inter-view Predicted Motion Candidate (IPMC). IPMC is introduced at the first place in the merge [46] candidate list, and the DV itself is inserted in the merge list as Disparity Motion Vector candidate (DMV). This description corresponds to the implementation of the HTM7.0 reference software, and modifications of these steps can occur during the further evolutions of the standard.

4.3.1.3 Improvements for Full Parallax

Full parallax SMV content can be encoded with a multi-view standard encoder with an adaptation of the inter-view references structure level. In [33], the views are first scanned in spiral as illustrated in Fig. 4.12a and realigned horizontally. The horizontal arrangement is then encoded by MVC with an IBP prediction structure (Fig. 4.12b). The resulting scheme of equivalent IBP structure with the views in two dimensions is illustrated in Fig. 4.12c, which shows that this approach is limited by the introduction of unsuitable and ineffective predictions.

In [47], horizontal IPP or IBP structures (Fig. 4.13e, f) are applied to each line of the views array, and a vertical inter-view prediction is added for the first or central column of views only, as illustrated in Fig. 4.14a–c. The main drawback of such structures is the limited number of available vertical inter-view predictions.

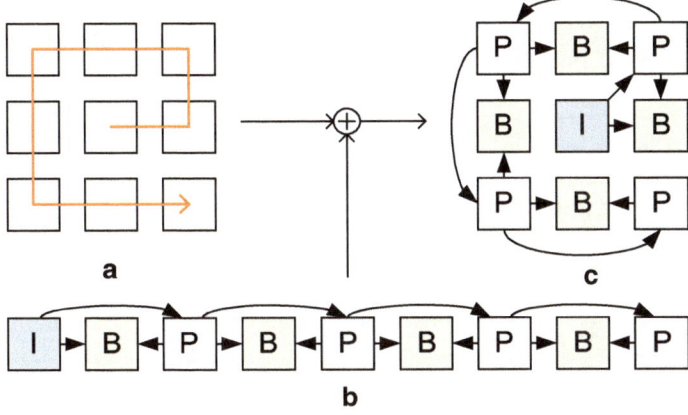

Fig. 4.12 State-of-the-art method [33] for 9 views (0...8) (**a**) spiral scan, (**b**) IBP structure for inter-view prediction, (**c**) equivalent IBP scheme in two dimensions

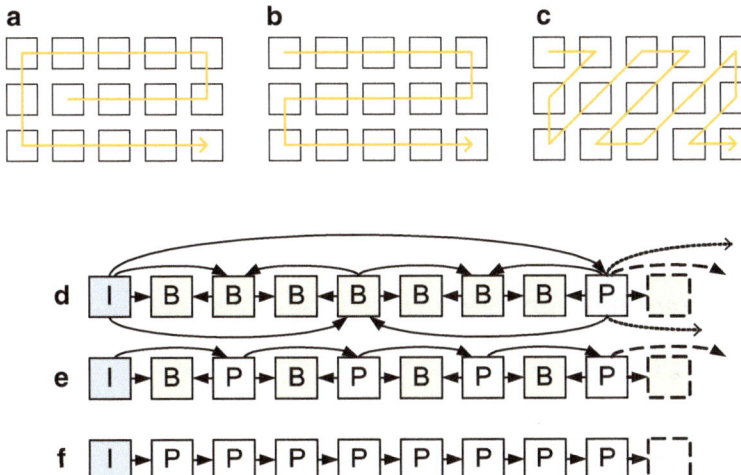

Fig. 4.13 Scan order: (**a**) spiral, (**b**) perpendicular, (**c**) diagonal and Horizontal inter-view reference picture structures: (**d**) hierarchical, (**e**) IBP, (**f**) IPP

In [48–50], the structure illustrated in Fig. 4.14d is proposed. Each line of views has IBP structure. Additional vertical inter-view predictions introduce views of types B1 with two horizontal or vertical only references, B2 with one horizontal and two vertical references, and B3 with two references in both directions. The small number of views that use both horizontal and vertical references (less than half of the views are of types B2 or B3) is the main limitation of this scheme, with the distance between the coding views and reference views.

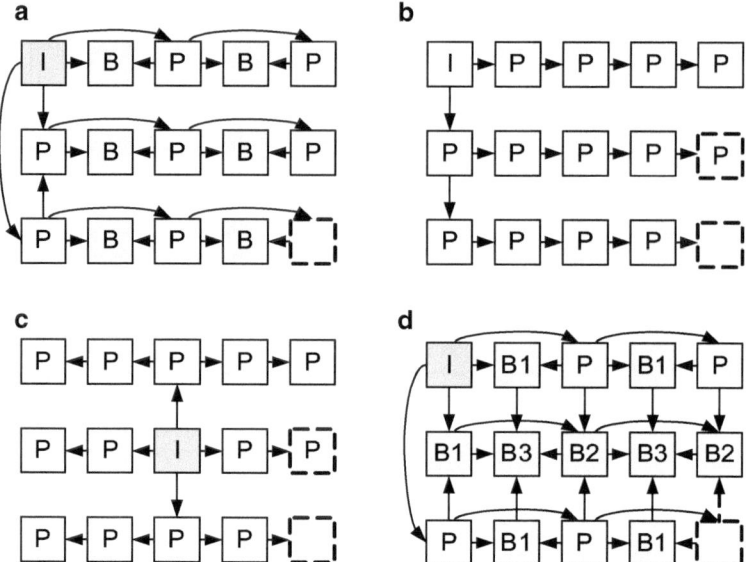

Fig. 4.14 State-of-the-art structures (**a**), (**b**), and (**c**) proposed in [47], and (**d**) proposed in [48–50]

In [48–51], an approach at the coding unit level is proposed, with several similar methods based on the prediction of a DV for the current coding view by interpolation of DVs from neighboring views.

4.3.2 Proposition for a New Inter-view Reference Pictures Configuration

4.3.2.1 Reference and Proposed Scheme

In [52] we propose the *Central2D* scheme illustrated in Fig. 4.15b. This two dimensional inter-view reference picture structure exploits the two dimensional view alignment. *Central2D* can be applied to an $N \times M$ views configuration. The central view is encoded first without inter-view references. Then the $N - 1$ (and respectively $M - 1$) views in the same horizontal (and resp. vertical) axis as the central view are encoded with only one inter-view reference (which is the nearest view in the central direction). Finally, the remaining views are encoded using one horizontal and one vertical inter-view references (which are the nearest views in the central direction). It enables the use of a horizontal and a vertical inter-view reference picture for a large number of views (only $M + N - 1$ views do not have horizontal and vertical reference pictures). This method minimizes the distance between the coding views and their inter-view reference pictures because it uses the closest neighbor and does not use diagonal references.

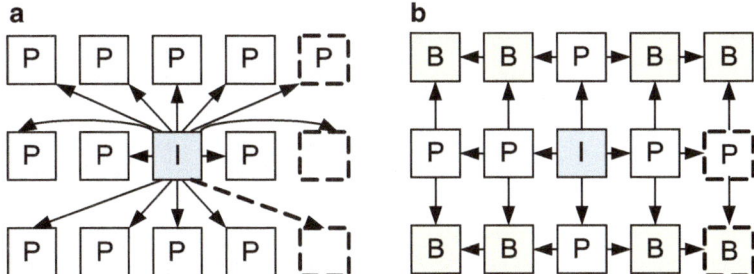

Fig. 4.15 (**a**) basic anchor, (**b**) proposed *Central2D*

Fig. 4.16 A frame of the
CoastalGuard sequence

In the following section, we assess the gain of efficiency provided by inter-view prediction in two directions and by a small distance between the coding and the reference views. The proposed *Central2D* structure is compared to a basic anchor with only the central view as inter-view reference picture for all the other views (illustrated in Fig. 4.15a). Our experiments also include state-of-the-art structures [47] and [48] illustrated in Fig. 4.14c, d, and the spiral scan with IBP structure [33] illustrated in Fig. 4.12. We also extend the method in [33] by varying the scan order and the structure as illustrated in Fig. 4.13.

4.3.2.2 Experimental Results

The state-of-the-art and proposed schemes are tested within MV-HEVC. Experiments are performed under MV-HEVC reference software version 7.0 (HTM7.0 with QC_MVHEVC macro). The temporal prediction structure remains as described in the Common Test Conditions (CTC) [53]. *CoastalGuard* (50 frames, computer generated, resolution 768×384) and *Akko&Kayo* (290 frames, captured, resolution 640×480) are the two sequences tested, with configurations of 3×3 views and 11×5 views (see Fig. 4.16). The results are measured using the Bjøntegaard Delta (BD) rate [54] over the QPs 22-27-32-37, with the basic anchor structure (Fig. 4.15a) as reference. We recall that the BD rate is the rate variation

Table 4.1 BD-rate variations for state-of-the-art and proposed structures compared to basic anchor—with 3 × 3 views

Coast 3 × 3

	Spiral	Perpendicular	Diagonal
IPP	−1.2 %	−2.2 %	5.1 %
IBP	9.1 %	7.1 %	11.4 %
Hierarchical	3.0 %	4.4 %	8.4 %
Method [48]	2.1 %		
Method [47]	−6.8 %		
CENTRAL2D	−7.1 %		

Akko 3 × 3

	Spiral	Perpendicular	Diagonal
IPP	−4.9 %	−5.5 %	8.8 %
IBP	2.7 %	−4.0 %	−1.9 %
Hierarchical	1.9 %	2.4 %	4.0 %
Method [48]	7.8 %		
Method [47]	−7.7 %		
CENTRAL2D	−8.2 %		

Table 4.2 BD-rate variations for state-of-the-art and proposed structures compared to basic anchor—with 11 × 5 views

Coast 11 × 5

	Spiral	Perpendicular	Diagonal
IPP	−20.5 %	−19.6 %	16.1 %
IBP	−15.9 %	−14.9 %	−13.9 %
Hierarchical	−8.4 %	−9.3 %	−13.0 %
Method [48]	−19.5 %		
Method [47]	−24.4 %		
CENTRAL2D	−29.1 %		

Akko 11 × 5

	Spiral	Perpendicular	Diagonal
IPP	−22.9 %	−24.8 %	−6.5 %
IBP	−20.0 %	−23.4 %	−2.4 %
Hierarchical	−14.9 %	−20.2 %	−3.7 %
Method [48]	−24.2 %		
Method [47]	−25.9 %		
CENTRAL2D	−27.6 %		

with respect to the reference at the same quality, hence negative values represent gains of the proposed technique.

Table 4.1 shows that *Central2D* scheme, method [47], and IPP structure with perpendicular and spiral scan outperform the other methods for both sequences with a 3 × 3 views configuration. These schemes minimize the distance between the coding views and the inter-view reference pictures and do not use diagonal inter-view reference pictures. *Central2D* has an additional gain due to the use of both horizontal and vertical inter-view reference pictures. Table 4.2 shows that *Central2D* is the most coherent and efficient configuration also with a larger number of views. The proposed structure *Central2D* provides final BD-rate gains up to 8.2 % and 29.1 % against the basic anchor, in the 3 × 3 and 11 × 5 views configuration, respectively.

4.3.3 Adaptation and Improvements of Inter-view Coding Tools

4.3.3.1 Merge Candidate List Improvement

A normative modification of NBDV and IVMP is also proposed in [52]. NBDV and IVMP coding tools are specifically implemented to work in the Common Test Conditions [53], with only one horizontal inter-view reference picture, which is the central baseview (with view index 0). First an adaptation is done that enables the use of several inter-view reference pictures, possibly horizontal or vertical, and with a view index different from 0 (not central).

In addition to this adaptation, we further improve NBDV. The modified NBDV searches for two DVs (one for each inter-view reference picture) when encoding a B view that uses one horizontal and one vertical inter-view reference pictures. This search for a second DV does not provide BD-rate gain in itself but is meant to be used for IPMC and DMV as follows. The new second DV allows to insert a second IPMC at the second place of the merge candidate list. Inter-view bi-prediction in both directions at the same time is enabled for the DMV merge candidate by using the couple of DVs.

4.3.3.2 Inter-view Derivation of the Second DV

In order to improve the efficiency of modified IPMC and DMV candidates, we propose to increase the likelihood of finding a second DV with NBDV. Figure 4.17 shows the steps of this method. For the current coding view, NBDV first finds a horizontal DV pointing to a reference CU in an inter-view reference picture. If this horizontal reference picture uses itself a vertical inter-view reference picture, and if the reference CU is coded by inter-view prediction, then the vertical DV used for the prediction is derived (i.e., copied) as a second DV for the current coding CU, and then used by IPMC and DMV as described in the previous section. This method can be applied to B views with one horizontal and one vertical references, which makes the *Central2D* structure the most adequate for these coding tools.

4.3.3.3 Experimental Results

The proposed adaptation and improvements of NBDV and IVMP coding tools are tested in this section. Experiments are performed under 3D-HEVC reference

Fig. 4.17 Inter-view
derivation of a second DV

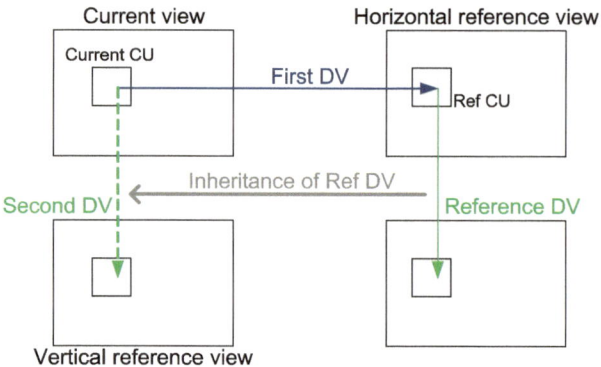

Table 4.3 BD-rate variations
for improved NBDV and
IVMP using one DV for each
inter-view reference picture

Reference: 3D-HEVC (HTM7.0 without modifications)

	3 × 3 views		11 × 5 views	
	Coast	Akko	Coast	Akko
Adaptation only	−1.1 %	−2.3 %	−2.4 %	−3.3 %
BiDMV	−1.2 %	−2.4 %	−2.7 %	−3.7 %
2 IPMC	−1.1 %	−2.3 %	−2.8 %	−3.5 %
Both	−1.3 %	−2.5 %	−3.1 %	−3.9 %

Table 4.4 BD-rate variations
for improved NBDV and
IVMP with inter-view
derivation of a second DV

Reference: 3D-HEVC (HTM7.0 without modifications)

	3 × 3 views		11 × 5 views	
	Coast	Akko	Coast	Akko
BiDMV + derivation	−1.9 %	−2.9 %	−3.4 %	−3.9 %
2 IPMC + derivation	−1.3 %	−2.4 %	−2.8 %	−3.5 %
Both + derivation	−2.0 %	−2.9 %	−3.9 %	−4.2 %

software version 7.0 (HTM7.0). The test conditions are the same as in Sect. 4.3.2.2,
(i.e., allowing two dimensional structures configuration). The *Central2D* structure
(proposed in Sect. 4.3.2) is used in all the following experiments. The reference is
HTM7.0 without software modifications.

In Table 4.3, the BD-rate gains provided by the adaptation of NBDV and IVMP
to a two dimensional structure are up to 3.3 %. It confirms the impact of the use of
horizontal and vertical dimensions at the inter-view references structure level. The
insertion of a second IPMC in the merge candidate list and the bi-prediction for the
DMV merge candidate separately increase the gains up to 2.4 % for the 3 × 3 views
configuration and 3.7 % for the 11 × 5 views configuration. The combination of
both provides a gain up to 2.5 % and 3.9 %, respectively, with 3 × 3 and with 11 ×
5 views. The combination of both tools provides slightly higher gain than the sum
of each taken separately because the bi-prediction allows NBDV to find more often
a second DV, which increases the chances to create a relevant second IPMC
candidate. Table 4.4 shows that the proposed derivation for the second DV is

efficient and increases the encoding performance of the complete proposed method (including the adaptation of NDBV and IVMP to a full parallax structure, the two IPMC, the DMV bi-prediction and the inter-view derivation of the second DV) up to 2.9 % and 4.2 % for the sequence *Akko&Kayo*, respectively, with 3×3 and with 11×5 views.

Conclusion

Current 3D video technologies are not able to provide the viewer with a smooth motion parallax, which is a very important cue in the perception of depth. Integral imaging and Super Multi-View video are 3D technologies that allow creating a light-field representation of a scene by capturing many viewpoints of that scene. The large number and the density of the views provide a smooth motion parallax in horizontal and potentially vertical dimensions.

However the large amount of data and its specific structure represent a challenge for future 3D video services and broadcasts. New efficient coding technologies are required. In the case of full parallax 3D content, several improvements are possible to exploit the vertical dimension. For integral imaging, many methods have been proposed based on 3D-DCT and wavelet transforms. Other methods propose to extract the views from an integral image and to encode them with standard encoders (e.g., HEVC) and their 3D extensions (e.g., MV-HEVC and 3D-HEVC). Moreover, improvements of these extensions have been proposed to encode content with horizontal and vertical parallax.

In this chapter, an inter-view reference picture structure adapted to 3D light-field video content with full motion parallax is proposed. Its main features are the use of both horizontal and vertical inter-view references, and the minimal distance between the coded and the reference views. The proposed structure outperforms a basic anchor by up to 29.1 % in BD-rate, showing the impact of the use of both horizontal and vertical directions in the inter-view reference picture scheme. Improvements of 3D-HEVC coding tools NBDV and IVMP are also proposed, in order to exploit both horizontal and vertical directions in a full parallax configuration, providing BD-rate gains up to 4.2 %. The results of the proposed methods show that exploiting efficiently both horizontal and vertical dimensions of full parallax SMV content at the coding structure and coding tools level can significantly improve the compression performance.

Acknowledgements The Coast sequence is provided by Orange Labs. The Akko&Kayo sequence is provided by Fujii Laboratory at Nagoya University.

References

1. Dufaux F, Pesquet-Popescu B, Cagnazzo M (2013) Emerging technologies for 3D video: content creation, coding, transmission and rendering. Wiley, New York
2. Tehrani MP, Senoh T, Okui M, Yamamoto K, Inoue N, Fujii T (2013) [m31095][FTV AHG] Use cases and application scenarios for super multiview video and free-navigation. In: International organisation for standardisation. ISO/IEC JTC1/SC29/WG11
3. Adelson EH, Bergen JR (1991) The plenoptic function and the elements of early vision. Computational models of visual processing, vol 1. MIT Press, Cambridge
4. Lumsdaine A, Georgiev T (2008) Full resolution lightfield rendering. Indiana University and Adobe Systems, Tech. Rep
5. Tehrani MP, Senoh T, Okui M, Yamamoto K, Inoue N, Fujii T (2013) [m31261][FTV AHG] Multiple aspects. In: International Organisation for Standardisation. ISO/IEC JTC1/SC29/WG11, October 2013
6. Tehrani MP, Senoh T, Okui M, Yamamoto K, Inoue N, Fujii T (2013) [m31103][FTV AHG] Introduction of super multiview video systems for requirement discussion. In: International Organisation for Standardisation. ISO/IEC JTC1/SC29/WG11, October 2013
7. Lippmann G (1908) Epreuves reversibles donnant la sensation du relief. J. Phys Theor Appl 7 (1):821–825
8. Georgiev T, Lumsdaine A (2010) Focused plenoptic camera and rendering. J Electron Imag 19 (2):021106
9. https://www.lytro.com/
10. https://raytrix.de/
11. http://www.tgeorgiev.net/
12. Iwasawa S, Kawakita M, Inoue N (2013) REI: an automultiscopic projection display. In: International conference on 3D systems and applications, vol 1, Osaka, June 2013
13. http://www.holografika.com/
14. http://gl.ict.usc.edu/Research/3DDisplay/
15. http://www.holymine3d.com/prod/prod03.html
16. Yendo T, Fujii T, Tanimoto M, Panahpour Tehrani M (2010) The Seelinder: Cylindrical 3D display viewable from 360 degrees. J Vis Commun Image Represent 21(5):586–594, 2010
17. Arai J, Kawai H, Okano F (2006) Microlens arrays for integral imaging system. Appl Opt 45 (36):9066–9078
18. Martinez-Cuenca R, Saavedra G, Martinez-Corral M, Javidi B (2009) Progress in 3-D multiperspective display by integral imaging. Proc. IEEE 97(6):1067–1077
19. Lino JFO (2013) 2D image rendering for 3D holoscopic content using disparity-assisted patch blending. In: Thesis to obtain the Master of Science Degree
20. Forman M, Aggoun A, McCormick M (1995) Compression of integral 3D TV pictures. In: Fifth international conference on image processing and its applications, IET, Edinburgh, July 1995, pp 584–588
21. Aggoun A (2006) A 3D DCT compression algorithm for omnidirectional integral images. In: 2006 I.E. international conference on acoustics, speech and signal processing. ICASSP 2006 Proceedings, vol. 2, IEEE, Toulouse, May 2006, pp II–II
22. Forman MC, Aggoun A, McCormick M (1997) A novel coding scheme for full parallax 3D-TV pictures. In: 1997 I.E. international conference on acoustics, speech, and signal processing. ICASSP 1997 Proceedings, vol. 4, IEEE, Nagoya, August 1997, pp 2945–2947
23. Forman M, Aggoun A (1997) Quantisation strategies for 3D-DCT-based compression of full parallax 3D images. In: Sixth international conference on image processing and its applications, IET, Dublin, January 1997, pp 32–35
24. Elharar E, Stern A, Hadar O, Javidi B (2007) A hybrid compression method for integral images using discrete wavelet transform and discrete cosine transform. J Disp Technol 3(3):321–325
25. Aggoun A (2011) Compression of 3D integral images using 3D wavelet transform. J Disp Technol 7(11):586–592

26. Kishk S, Ahmed HEM, Helmy H (2011) Integral images compression using discrete wavelets and PCA. Int J Signal Process Image Process Pattern Recognit 4(2):65–78
27. Zayed HH, Kishk SE, Ahmed HM (2012) 3D wavelets with SPIHT coding for integral imaging compression. Int J Comput Sci Netw Secur 12(1):126
28. Olsson R, Sjostrom M, Xu Y (2006) A combined pre-processing and H. 264 compression scheme for 3D integral images. In: 2006 I.E. international conference on image processing, IEEE, Atlanta, October 2006, pp 513–516
29. Yeom S, Stern A, Javidi B (2004) Compression of 3D color integral images. Opt Express 12 (8):1632–1642
30. Yan P, Xianyuan Y (2011) Integral image compression based on optical characteristic. IET Comput Vis 5(3):164–168
31. Kang H-H, Shin D-H, E-S Kim (2008) Compression scheme of sub-images using karhunen-loeve transform in three-dimensional integral imaging. Opt Commun 281(14):3640–3647
32. Dick J, Almeida H, Soares LD, Nunes P (2011) 3D holoscopic video coding using MVC. In: 2011 I.E. EUROCON-international conference on computer as a tool, IEEE, Lisbon, April 2011, pp 1–4
33. Shi S, Gioia P, Madec G (2011) Efficient compression method for integral images using multi-view video coding. In: 18th IEEE international conference on image processing (ICIP), IEEE, Brussels, September 2011, pp 137–140
34. Ohm J-R (2013) Overview of 3D video coding standardization. In: International conference on 3D systems and applications, Osaka, June 2013
35. Adedoyin S, Fernando WAC, Aggoun A (2007) A joint motion & disparity motion estimation technique for 3D integral video compression using evolutionary strategy. IEEE Trans Consum Electron 53(2):732–739
36. Adedoyin S, Fernando WAC, Aggoun A, Kondoz K (2007) Motion and disparity estimation with self adapted evolutionary strategy in 3D video coding. IEEE Trans Consum Electron 53 (4):1768–1775
37. Marpe D, Wiegand T, Sullivan GJ (2006) The H. 264/MPEG4 advanced video coding standard and its applications. Commun Mag IEEE 44(8):134–143
38. Sullivan GJ, Ohm J, Han W-J, Wiegand T (2012) Overview of the high efficiency video coding (HEVC) standard. IEEE Trans Circuits Syst Video Technol 22(12):1649–1668
39. Conti C, Lino J, Nunes P, Soares LD, Correia PL (2011) Improved spatial prediction for 3D holoscopic image and video coding. In Proc. EURASIP european signal processing conference (EUSIPCO), Barcelona, August 2011
40. Conti C, Lino J, Nunes P, Soares LD, Lobato Correia P (2011) Spatial prediction based on self-similarity compensation for 3D holoscopic image and video coding. In: 18th IEEE international conference on image processing (ICIP), IEEE, Brussels, September 2011, pp 961–964
41. Conti C, Nunes P, Soares LD (2012) New HEVC prediction modes for 3D holoscopic video coding. In: 19th IEEE international conference on image processing (ICIP), IEEE, Orlando, September–October 2012, pp 1325–1328
42. Conti C, Lino J, Nunes P, Soares LD (2012) Spatial and temporal prediction scheme for 3D holoscopic video coding based on H. 264/AVC. In: 19th international packet video workshop (PV), IEEE, Munich, May 2012, pp 143–148
43. Conti C, Nunes P, Soares LD (2013) Using self-similarity compensation for improving inter-layer prediction in scalable 3D holoscopic video coding. In: SPIE optical engineering and applications. International society for optics and photonics, September 2013, pp 88 561K–88 561K
44. Zhang L, Chen Y, Karczewicz M (2012) 3D-CE5.h related: Disparity vector derivation for multiview video and 3DV. In: International Organisation for Standardisation. ISO/IEC JTC1/SC29/WG11 MPEG2012/m24937, July 2012
45. Tech G, Wegner K, Chen Y, Yea S (2012) 3D-HEVC test model 2. In: International Organisation for Standardisation. ITU-T SG 16 WP 3 & ISO/IEC JTC1/SC29/WG11 JCT3V-B1005, October 2012

46. Helle P, Oudin S, Bross B, Marpe D, Bici MO, Ugur K, Jung J, Clare G, Wiegand T (2012) Block merging for quadtree-based partitioning in hevc. IEEE Trans Circuits Syst Video Technol 22(12):1720–1731
47. Merkle P, Smolic A, Muller K, Wiegand T (2007) Efficient prediction structures for multiview video coding. IEEE Trans Circuits Syst Video Technol 17(11):1461–1473
48. Chung T, Song K, Kim C-S (2008) Compression of 2-D wide multi-view video sequences using view interpolation. In: 15th IEEE international conference on image processing (ICIP), IEEE, San Diego, October 2008, pp 2440–2443
49. Chung T-Y, Jung I-L, Song K, Kim C-S (2009) Virtual view interpolation and prediction structure for full parallax multi-view video. In: Advances in multimedia information processing - PCM, vol 5879. Springer, Berlin, pp 543–550
50. Chung T-Y, Jung I-L, Song K, Kim C-S (2010) Multi-view video coding with view interpolation prediction for 2D camera arrays. J Vis Commun Image Represent 21(5):474–486
51. Avci A, De Cock J, Lambert P, Beernaert R, De Smet J, Bogaert L, Meuret Y, Thienpont H, De Smet H (2012) Efficient disparity vector prediction schemes with modified P frame for 2D camera arrays. J Vis Commun Image Represent 23(2):287–292
52. Dricot A, Jung J, Cagnazzo M, Pesquet B, Dufaux F (2014) Full parallax super multi-view video coding. In: 21st IEEE International Conference on Image Processing (ICIP), IEEE, Paris, October 2014
53. Rusanovsky D, Muller K, Vetro A (2012) Common test conditions of 3DV core experiments. In: International Organisation for Standardisation. ITU-T SG 16 WP 3 & ISO/IEC JTC1/SC29/WG11 JCT3V-B11000, October 2012
54. Bjøntegaard G (2001) Calculation of average PSNR differences between RD-curves. In: VCEG Meeting, Austin, April 2001

Chapter 5
3D Holoscopic Video Representation and Coding Technology

Caroline Conti, Luís Ducla Soares, and Paulo Nunes

Abstract 3D holoscopic imaging, also known as integral imaging, light-field imaging, or plenoptic imaging, has been attracting the attention of the research community as a prospective technology for 3D acquisition and visualization. However, to make this technology a commercially viable candidate for three-dimensional services, there are several important requirements that still need to be fulfilled. This chapter provides an overview of some of these critical success factors with special emphasis on the need for a suitable representation format in conjunction with an efficient coding scheme for 3D holoscopic content. Moreover, an efficient 3D holoscopic coding solution is described, which provides backward compatibility with legacy 2D and 3D multiview displays by using a multi-layer scalable coding architecture. In addition to this, novel prediction methods are presented to enable efficient storage and delivery of 3D holoscopic content to the end-users.

5.1 Introduction

Motivated by the notable success of three-dimensional (3D) movies in the last decade, 3D display technologies have been extensively researched and there has been a significant effort for promoting the adoption of three-dimensional television (3DTV).

Currently, the most used 3D display technology is based on the stereoscopic approach, where the 3D sensation is induced by presenting two different views to the user, one for each eye of the observer, through some kind of multiplexing (e.g., spatial or temporal multiplexing). Although 3D-capable television sets based on this technique are increasingly available, their public acceptance is far from what was expected by the industry. One of the main reasons for this is the need for wearing special glasses to induce the 3D sensation, which is not suitable for a living room environment and may cause discomfort for some users.

C. Conti (✉) • L.D. Soares • P. Nunes
ISCTE - Instituto Universitário de Lisboa/Instituto de Telecomunicações, Lisbon, Portugal
e-mail: caroline.conti@lx.it.pt; lds@lx.it.pt; paulo.nunes@lx.it.pt

© Springer Science+Business Media New York 2015
A. Kondoz, T. Dagiuklas (eds.), *Novel 3D Media Technologies*,
DOI 10.1007/978-1-4939-2026-6_5

On the other hand, autostereoscopic techniques, which do not require any special eyewear, have also been brought up as candidates for future 3D display solutions. Among these solutions, the most likely to appear in a second step of 3DTV development is the multiview approach, which allows seeing a discrete number of perspectives of the scene with horizontal parallax. However, both stereoscopic and multiview approaches can lead to considerable visual discomfort that may inhibit their usage for applications that require prolonged observation periods.

In this context, 3D holoscopic imaging, also known as integral, light-field, and plenoptic imaging, has become an attractive alternative for providing 3D content and enabling a fatigue-free 3D visualization. Furthermore, following the recent improvements in micro-lenses manufacturing, image sensors, and flat display devices, 3D holoscopic imaging is turning into a practical technology and promises to become popular in the future.

Besides the aforementioned advantages of 3D holoscopic imaging, there are several critical success factors which need to be considered in order to introduce this technology into the consumer electronics market for 3DTV applications, notably: coding efficiency, bandwidth requirements, receiver types (fixed, mobile, etc.), quality of experience, as well as backward compatibility with legacy 2D devices. These factors are closely related to all stages of the 3DTV chain—acquisition, representation, coding, transmission, reception, and display—and are analyzed in this chapter with special emphasis on the representation and coding stages, so as to efficiently deliver 3D holoscopic content to the end-users.

As an insightful example, this chapter presents a display scalable coding scheme for 3D holoscopic content. In this scheme, a multi-layer display scalable architecture is used, where each layer represents a different level of display scalability: Layer 0—a single 2D view; Layer 1—3D stereo or multiview; and Layer 2—the full 3D holoscopic content. Furthermore, novel prediction methods are used to improve the coding efficiency by taking advantage of the redundancy existing between multiview and 3D holoscopic content, as well as of the inherent correlation of the 3D holoscopic content.

The remainder of this chapter is organized as follows: Section 5.2 overviews the principles behind the 3D holoscopic imaging approach and discusses quality of experience issues that still need to be overcome. Section 5.3 presents possible representation formats for 3D holoscopic content and also reviews some algorithms that can be used to generate 2D and multiview from 3D holoscopic content. Section 5.4 describes the display scalable coding solution for 3D holoscopic content. Section 5.5 presents some experimental results and, finally, Section 5.6 concludes the chapter.

5.2   Acquisition and Display Systems for 3D Holoscopic Content

3D holoscopic imaging derives from the fundamental concepts of light-field imaging where, by using conventional 2D acquisition and display devices enhanced by special optical systems, it is possible to capture not only spatial information about a scene but also angular information, i.e., the "whole observable" (holoscopic) scene. This section provides an overview of the principles behind 3D holoscopic imaging and discusses the advantages and current limitations of this approach for acquisition and display.

5.2.1   3D Holoscopic Imaging Acquisition

The concept behind 3D holoscopic imaging is based on the principle of integral photography, proposed by Gabriel Lippmman in 1908 [1]. Basically, the 3D holoscopic imaging system comprises a regularly spaced array of small spherical micro-lenses, known as a "fly's eye" lens array [2], which is used for both acquisition and display of the 3D holoscopic images, as can be seen in Fig. 5.1.

At the acquisition side (Fig. 5.1a), the light beams coming from a given object with various incident angles are firstly refracted through the micro-lens array to be, then, captured by the 2D image sensor. Hence, each micro-lens works as an individual small low resolution camera conveying a particular perspective of the 3D object at slightly different angles. As a result, the planar light intensity distribution representing a 3D holoscopic image corresponds to a 2D array of micro-images, as illustrated in Fig. 5.2, where both light intensity and directional information are recorded.

Apart from some technological limitations that still need to be overcome in this kind of systems [2], there have been several camera setups seeking to improve the performance of this technology, as can be seen in [2–4]. Moreover, various advantages of employing a 3D holoscopic system for the acquisition of 3D content have been identified, notably:

- A single aperture camera system is used, without needing complex calibration and synchronization procedures between several cameras;
- A high level of display scalability is supported, since it is possible to generate in a seamless way 2D, multiview, as well as full 3D holoscopic, from a 3D holoscopic acquisition setup (see Sect. 5.3.4);
- New degrees of freedom for content production become possible at the post-processing stage, such as refocusing and changing perspectives (see Sect. 5.3.4).

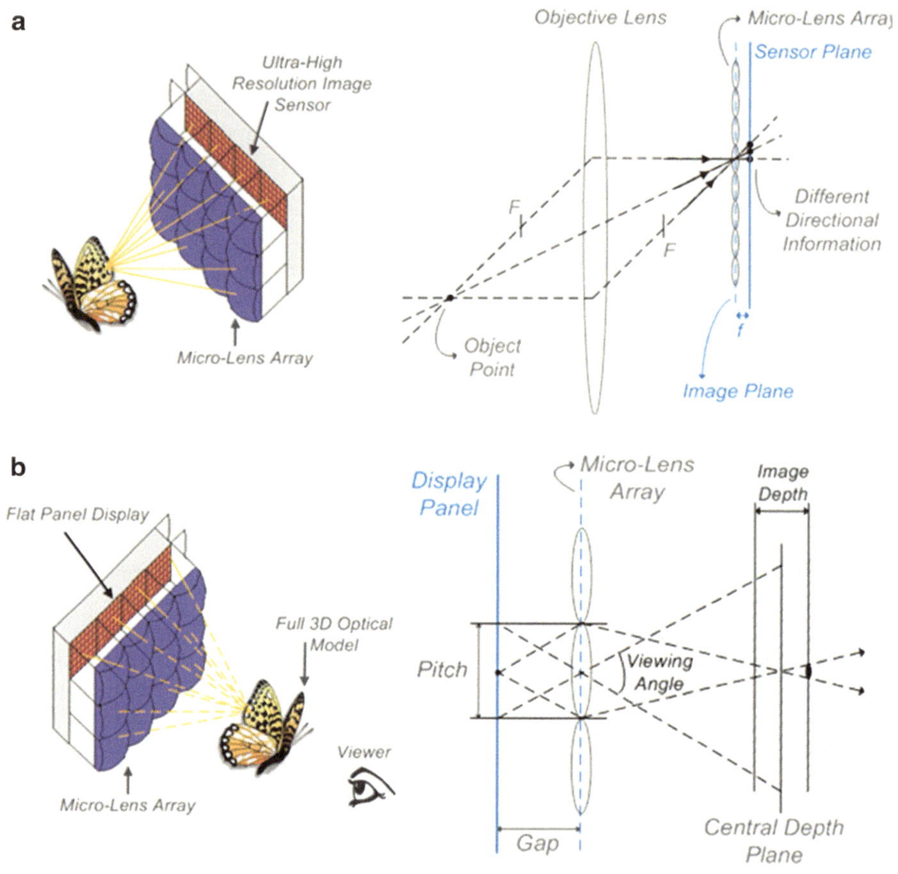

a

Ultra-High
Resolution Image
Sensor

Micro-Lens Array

Objective Lens

Micro-Lens Array

Sensor Plane

F

F

Different
Directional
Information

Object
Point

f

Image Plane

b

Flat Panel Display

Full 3D Optical
Model

Viewer

Micro-Lens Array

Display
Panel

Micro-Lens
Array

Image
Depth

Pitch

Viewing
Angle

Gap

Central Depth
Plane

Fig. 5.1 A 3D holoscopic imaging system: (**a**) acquisition side; (**b**) display side

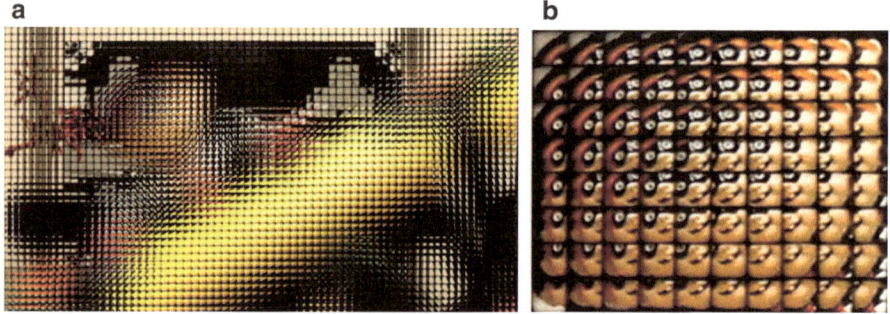

a

b

Fig. 5.2 3D holoscopic image captured with a 250 μm pitch micro-lens array: (**a**) full image with resolution of 1920×1088; (**b**) enlargement of 280×224 pixels showing the array of micro-images

5.2.2 3D Holoscopic Imaging Display

For displaying 3D holoscopic content, a similar optical setup can be used, where a simple flat panel display is used to project the recorded 3D holoscopic image through the micro-lens array, as illustrated in Fig. 5.1b. Then, the intersection of the light beams passing through the micro-lenses recreates a full 3D optical model of the captured object, which is observable without the need for special glasses.

3D holoscopic imaging is one of the promising autostereoscopic techniques for improving the quality of experience in 3D visualization. Due to the used optical arrangement, a 3D holoscopic imaging display presents the following advantages compared to current stereo and multiview displays:

- An immersive 3D experience for more than one user (simultaneously) is provided;
- Continuous motion parallax in the horizontal and vertical directions (throughout the viewing zone) is made possible; and
- A more natural and fatigue-free 3D sensation with accurate convergence/accommodation and depth perception is offered to the user.

However, a critical issue, at this time, is related to the balance between the various viewing parameters (see Fig. 5.1b) of a 3D holoscopic display system. As explained in [5], the quality of the reconstructed 3D image in a 3D holoscopic display can be analyzed in terms of the following viewing parameters:

- Image Depth: Image depth refers to the distance between the nearest and farthest plane where the reconstructed 3D image can be observed without severe artifacts [5] (appearing with acceptable quality), as illustrated in Fig. 5.1b. To increase the image depth, it is necessary to reduce the pitch of the micro-lens (see Fig. 5.1b) and to increase the pixel size of the display. However, these conditions lead to a narrower viewing angle and smaller viewing resolution, as explained below.
- Viewing Angle: The viewing angle corresponds to the angular region where the reconstructed 3D image (recreated by the display) can be seen without flipping [5]. The viewing angle, as depicted in Fig. 5.1b, determines the area (so-called viewing zone) where the reconstructed 3D image can be observed with full motion parallax. Moreover, the number of possible different viewpoint images is determined by the number of pixels behind each micro-lens. As can be seen in Fig. 5.1b, the viewing angle is intrinsically limited by the pitch of the micro-lens and the gap between the sensor panel and the micro-lens array. Namely, the viewing angle increases as the pitch increases and the gap decreases. However, a larger micro-lens pitch leads to a smaller image depth as well as a smaller viewing resolution (as explained below).
- Viewing Resolution: The viewing resolution is the resolution of the reconstructed 3D image (visible to the user), which is essentially limited by the resolution of the display device and the pitch of the micro-lens. Considering a fixed display resolution, the pitch of the micro-lens determines the fraction of

the total display resolution which is sampled for angular direction information. Generally, the viewing resolution is simply defined by the number of micro-lenses in the array. Hence, it is possible to enhance the viewing resolution by decreasing the pitch of the micro-lens. However, a smaller micro-lens pitch leads to a narrower viewing angle.

Therefore, an important requirement to improve one viewing parameter without degrading the others (i.e., to provide 3D holoscopic content with good visual quality) is to have acquisition and display devices with very high resolutions. Consequently, efficient coding solutions become of cardinal significance to deal with the large amount of information involved.

5.3   3D Holoscopic Video Representation Formats

A key requirement for successful 3D video applications is the choice of an efficient representation in conjunction with an efficient coding scheme. In this sense, this section describes three possible formats for representing the 3D holoscopic image data: *micro-image based, viewpoint-image based,* and *ray-space-image based.* It will be seen, in Sect. 5.5.1, that a major factor for choosing a representation format over another may be the compression efficiency it allows. Notice that, a depth-enhanced representation of the 3D holoscopic content is also possible, but is not addressed in this chapter. Moreover, to the best of authors' knowledge, depth-enhanced 3D holoscopic representation and coding solutions have not been proposed in the literature yet.

Furthermore, as discussed in Sect. 5.2.1, one of the advantages of using a 3D holoscopic imaging acquisition system is its high-level of display scalability, being able, for instance, to generate 2D, stereoscopic 3D, and multiview content. Several algorithms to generate 2D content from a 3D holoscopic image have been proposed in the literature, mainly in the context of richer 2D image capturing systems [6]. In Sect. 5.3.4, two of these algorithms, referred to as Basic Rendering and Weighted Blending [7], are briefly reviewed, since they are used to generate the content for each hierarchical layer of the scalable 3D holoscopic coding architecture described later in Sect. 5.4.

5.3.1   Micro-Image Based Representation

As a result of the used optical acquisition setup, where a micro-lens array is overlapped with the image sensor, the 3D holoscopic image is inherently composed of a 2D array of slightly different micro-images. Due to the different nature of the 3D holoscopic content, the first step to build an efficient coding scheme for this type of content is to characterize their intrinsic correlations.

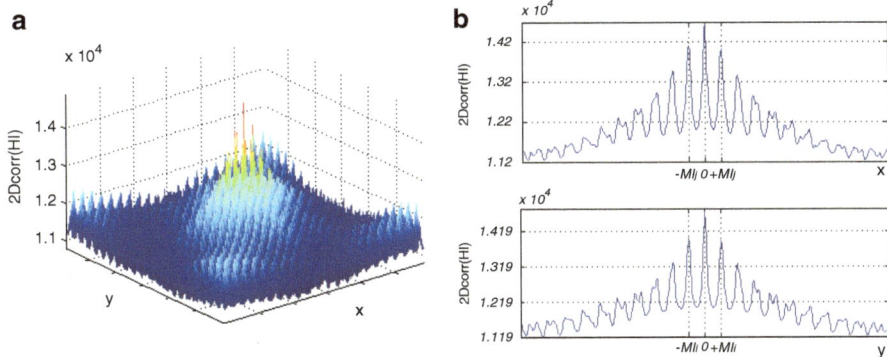

Fig. 5.3 Example of inherent 3D holoscopic spatial correlation: (**a**) autocorrelation function for a 3D holoscopic image; and (**b**) projection onto x and y axis, showing the high correlation between points spaced of about one micro-image size (MI)

Although there is a resemblance between 3D holoscopic video and 2D video (as both are captured by an ordinary 2D sensor), a more careful analysis reveals inherent correlations that are not exploited by state-of-the-art 2D video coding solutions. Notably, in the spatial domain, a significant correlation between neighboring micro-images can be identified through the autocorrelation function, as illustrated in Fig. 5.3. In particular, it can be seen that the pixel correlation in 3D holoscopic content is not as smooth as in conventional 2D video content. A periodic structure is evidenced by the autocorrelation function whose period is approximately one micro-image size (represented in each direction by MI_j and MI_i in Fig. 5.3b). Additionally, it should also be noted that each micro-image itself has some degree of inter-pixel redundancy, as is also common in 2D images.

Some coding schemes in the literature have proposed to represent the 3D holoscopic image by a stack of their composing micro-images, which can be interpreted as a pseudo volumetric image (PVI) [8, 9]—if stacking micro-images are done along the third dimension—or a pseudo video sequence (PVS) [10, 11]—if stacking is done along the temporal dimension.

Briefly, consider a 3D holoscopic image, HI, as illustrated in Fig. 5.4a, captured using a rectangular-packed square-based micro-lens array with resolution $MLA_n \times MLA_m$ and micro-image resolution of $MI_j \times MI_i$. Each micro-image, $MI_\mathbf{k}$, in the PVI or PVS representation can be obtained from HI at the position $\mathbf{k} = (k_n, k_m)$ in the micro-lens array, as given by (5.1), where $\mathbf{x} = (x, y)$ represents the pixel positions inside $MI_\mathbf{k}$. Alternatively, the holoscopic image can be expressed in terms of its micro-images by the array in (5.2).

$$MI_\mathbf{k} = HI\left(k_n \times MI_j + x, \ k_m \times MI_i + y\right) \tag{5.1}$$

$$HI = [MI_\mathbf{k}]_{MLA_n \times MLA_m} \tag{5.2}$$

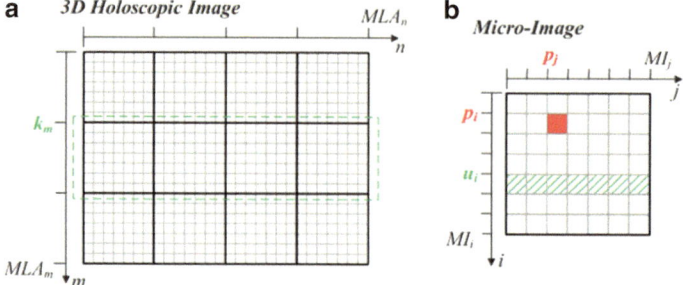

Fig. 5.4 Process to construct each representations format of the 3D holoscopic content

5.3.2 Viewpoint Image Based Representation

A viewpoint image based representation can be constructed by extracting one pixel (with the same relative position) from all micro-images of a given 3D holoscopic image. Hence, each resulting viewpoint image represents an orthographic projection of the complete 3D scene in a particular direction.

For example, considering the same 3D holoscopic image HI illustrated in Fig. 5.4a, a viewpoint image, $VI_\mathbf{p}$, is formed by taking one pixel at a fixed relative position $\mathbf{p} = (p_i , p_j)$ (see Fig. 5.4b) from each micro-image (in the array of $MLA_n \times MLA_m$ micro-images). Hence, the viewpoint image $VI_\mathbf{p}$ is given by (5.3), where $\mathbf{x} = (x, y)$ represents the pixel positions inside $VI_\mathbf{p}$. Consequently, the resolution of $VI_\mathbf{p}$ is given by the resolution of the micro-lens array, i.e., $MLA_n \times MLA_m$.

$$VI_\mathbf{p} = HI\big(x \times MI_j + p_i, \; y \times MI_i + p_j\big) \qquad (5.3)$$

Based on this type of representation format, some coding approaches in the literature proposed to exploit the existing disparity between adjacent viewpoint images by stacking these viewpoint images as PVIs [12] and PVSs [10, 13].

A possible alternative representation is to build a new 2D image from all viewpoint images. This new image, hereinafter referred to as VI-based holoscopic image (HI^{VI}), has the same resolution as the original 3D holoscopic image HI and can be constructed by arranging each $VI_\mathbf{p}$ according to the position given by $\mathbf{p}$. The VI-based holoscopic image is then composed of an array of $MI_j \times MI_i$ viewpoint images, i.e.,

$$HI^{VI} = \big[VI_\mathbf{p}\big]_{MI_j \times MI_i} \qquad (5.4)$$

Figure 5.5a shows an example of a VI-based holoscopic image. It is possible to see in the enlargement (Fig. 5.5b) that this image also has a repetitive structure, which may be likely exploited, similarly to the micro-image based representation. As can

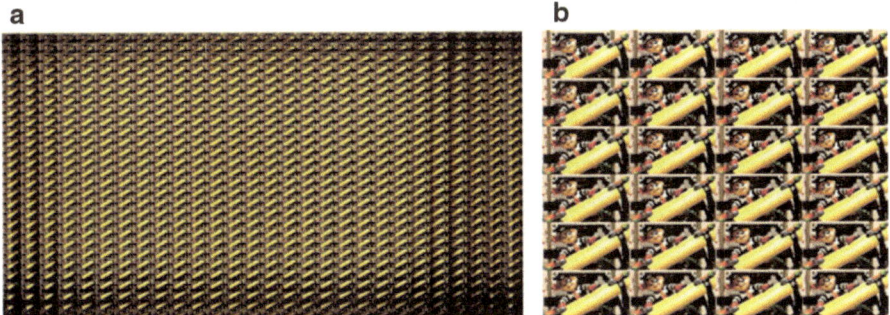

Fig. 5.5 Viewpoint image based representation: (**a**) VI-based holoscopic image; (**b**) enlargement of 272×228 pixels showing each viewpoint image in detail

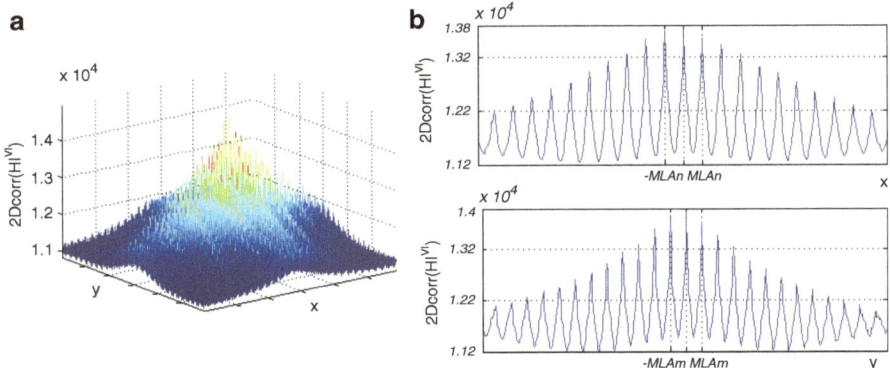

Fig. 5.6 Example of spatial correlation in a VI-based holoscopic image: (**a**) autocorrelation function; and (**b**) projection onto x and y axis

be seen in Fig. 5.6 through the autocorrelation function, the pixel correlation in the VI-based holoscopic image also presents a periodic structure, which is given approximately by the resolution of each viewpoint image (represented by $\text{MLA}_n \times \text{MLA}_m$ in Fig. 5.6b).

5.3.3   Ray-Space Image Based Representation

The epipolar-plane technique [14] can be used to generate a ray-space image based representation from the 3D holoscopic image. In this case, a ray-space image can be formed by stacking together micro-images in the same row/column (in the array of micro-lenses) and, then, taking a slice from a particular horizontal plane (perpendicular to the micro-lens plane). For example, choosing a row k_m in the array of micro-lenses illustrated in Fig. 5.4a, and considering that the horizontal plane was extracted from a row u_i inside the micro-image (see Fig. 5.4b), it is possible to define a row vector, $\mathbf{r}$, in the holoscopic image HI, which is given by (5.5).

$$\mathbf{r}(m = k_m, i = u_i) = k_m \times \mathrm{MI}_i + u_i \tag{5.5}$$

Thus, for a fixed row vector $\mathbf{r}(m, i)$, the corresponding ray-space image, $\mathrm{RI}_\mathbf{r}$, can be given by (5.6), where $\mathbf{x} = (x, y)$ corresponds to the relative pixel position inside this ray-space image. Consequently, the resolution of $\mathrm{RI}_\mathbf{r}$ is, then, $\mathrm{MLA}_n \times \mathrm{MI}_j$.

$$\mathrm{RI}_\mathbf{r}(\mathbf{x}) = \mathrm{HI}\big(\mathbf{r}, \ y \times \mathrm{MI}_j + x\big) \tag{5.6}$$

Similarly to the VI-based holoscopic image ($\mathrm{HI}^{\mathrm{VI}}$), an RI-based holoscopic image, $\mathrm{HI}^{\mathrm{RI}}$, can also be defined, which has the same resolution of the original 3D holoscopic image HI if (and only if) the micro-images are square. Considering square micro-images (i.e., $\mathrm{MI}_j = \mathrm{MI}_i$), this is possible by transposing each extracted $\mathrm{RI}_\mathbf{r}$ and arranging it at the relative position given by $\mathbf{r}$. In other words,

$$\mathrm{HI}^{\mathrm{RI}} = \big[\mathrm{RI}_\mathbf{r}^{\ \mathrm{T}}\big]_{\mathrm{MLA}_m \times \mathrm{MI}_i} \tag{5.7}$$

Through this format, the angular information in one direction (horizontal in this case) across different micro-images is represented. Figure 5.7a shows an example of the referred RI-based holoscopic image. It is possible to see in the enlargement (Fig. 5.7b) that there is also a significant amount of redundancy between neighboring ray-space images. However, as can be seen in Fig. 5.8 with the autocorrelation function, this redundancy is not equally distributed in horizontal and vertical directions. Additionally, the periodic structure is also observed, as seen in Fig. 5.8b, and is given, approximately, by the resolution in each direction (represented by $\mathrm{MLA}_n \times \mathrm{MI}_j$ in Fig. 5.8b).

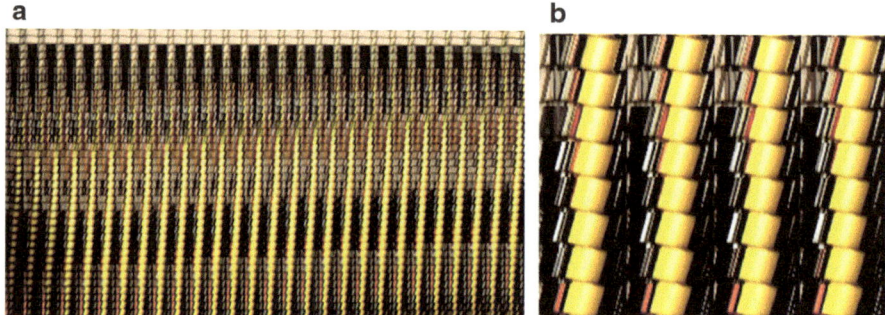

Fig. 5.7 Ray-space image based representation: (**a**) RI-based holoscopic image; (**b**) enlargement of 272 × 224 pixels showing each ray-space image in detail

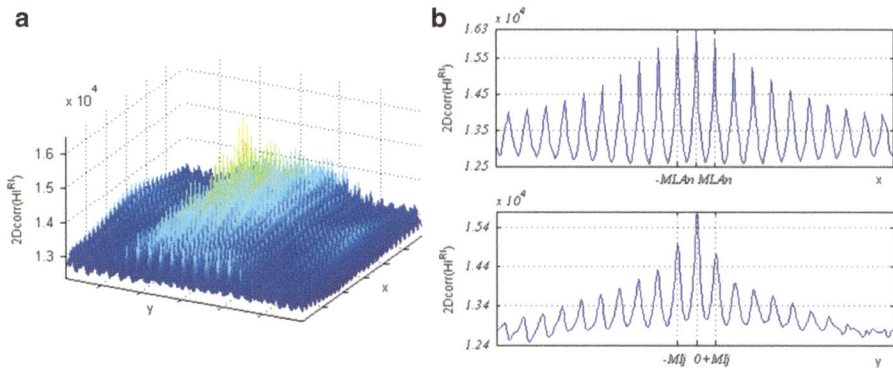

Fig. 5.8 Example of spatial correlation in a RI-based holoscopic image: (**a**) autocorrelation function; and (**b**) projection onto x and y axis

5.3.4   2D and Multiview Content Generation

Generating 2D and 3D multiview content from 3D holoscopic content means producing various 2D views with different viewing angles. Two algorithms proposed in [7] and referred to as Basic Rendering and Weighted Blending are briefly described, as follows. More details can be also found in [7].

5.3.4.1   Basic Rendering Algorithm

Since each micro-image can be seen as a low resolution view of the scene, it is possible to choose suitable portions from each micro-image to stitch and then compose a 2D view image. This is the basis for the Basic Rendering algorithm.

The input for this algorithm is a 3D holoscopic image (Fig. 5.9a) with an $MLA_n \times MLA_m$ array of micro-images, where each micro-image has a resolution of $MI_j \times MI_i$. In this 3D holoscopic image, a portion (patch) of $PS \times PS$ pixels is extracted from each micro-image. These patches are then stitched together, as illustrated in Fig. 5.9b. As a result, the output is a 2D view image with a resolution of $(PS \times MLA_n) \times (PS \times MLA_m)$.

5.3.4.2   Weighted Blending Algorithm

To avoid some blocky artifacts due to a non-perfect match between the patches with fixed size in the Basic Rendering algorithm, an alternative algorithm was proposed in [7] and is also considered here for 2D views generation. The main idea is to smooth these artifacts with a weighted blending algorithm.

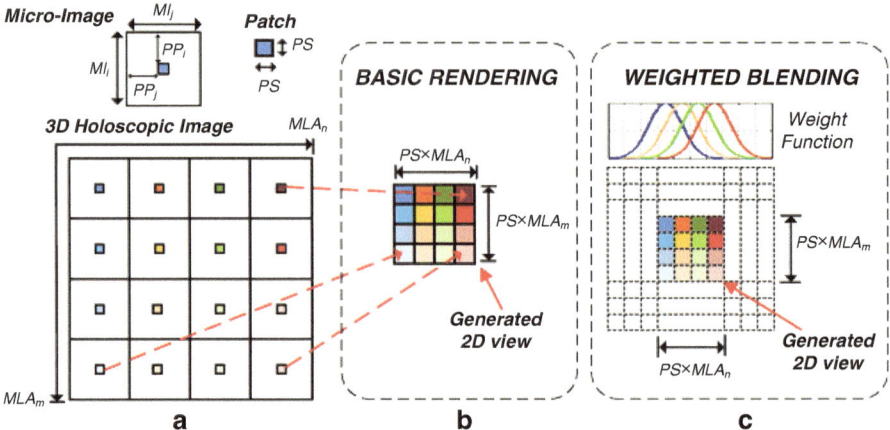

Fig. 5.9 Generating a 2D view image [15] (Copyright © 2013, IEEE): (**a**) 3D holoscopic input image; (**b**) Basic Rendering algorithm [7]; and (**c**) Weighted Blending algorithm [7]

This algorithm makes use of the knowledge that there is redundant spatial information in multiple micro-images, since each micro-image captures overlapping areas of the scene. Therefore, the weighted blending consists in averaging together all these overlapping regions across different micro-images, weighting differently the overlapping pixels. As can be seen in Fig. 5.9c, each micro-image is overlapped with a shift of PS pixels (corresponding to the patch size) to its neighboring micro-images in both horizontal and vertical directions. Then the pixels in the same spatial position across various micro-images are averaged by using a specific weighting function. This weighting function corresponds to a bivariate Gaussian function (non-normalized) of size $MI_j \times MI_i$ and whose mean vector, μ, is determined by the patch center, and the covariance matrix, Σ, is defined depending on the patch size. Hence, the weight applied to a pixel in the position $\mathbf{p} = (p_i\ ,\ p_j)$ inside its corresponding micro-image is given by (5.8). Similarly to the basic algorithm, by varying the position of the patch, it is possible to extract several different 2D view images, each having a $(PS \times MLA_n) \times (PS \times MLA_m)$ resolution.

$$\text{Weight}(\mathbf{p}) = \text{Gaussian}\left(\mathbf{p}|\mu,\ \Sigma\right) = \alpha\exp\left(\frac{1}{2}(\mathbf{p} - \mu)^T\Sigma^{-1}(\mathbf{p} - \mu)\right) \qquad (5.8)$$

5.3.4.3   Common Parameters

The Basic Rendering and Weighted Blending algorithms can be controlled through the following two main parameters:

1. *Size of the patch* (PS): Since a micro-image is captured by the corresponding micro-lens in perspective projection geometry, placing two objects of the same (real) size closer or farther from the micro-lens array will result in those objects

appearing with greater or smaller size in pixels in the various micro-images. Thus, for one of these two objects to appear sharp in the generated 2D view image, different patch sizes need to be selected (with larger and smaller sizes, respectively) from each micro-image. This fact is explained in more detail in [7], where it is also shown that it is possible to control the plane of focus in the generated 2D view image (i.e., which objects will appear in sharp focus) by choosing a suitable patch size. An important issue with these algorithms is that the resolution of the output 2D view depends on the selected plane of focus and, thus, on the patch size.

2. *Position of the patch* (PP): By varying the relative position of the patch in the micro-image, it is possible to generate 2D views with different horizontal and vertical viewing angles (i.e., different scene perspectives). Since the 3D multiview content usually represents perspectives with different horizontal angles of projection, 3D multiview content can be generated by varying the horizontal position of the patch (relative to the center of the micro-image).

5.4 3D Holoscopic Video Coding

As discussed in Sect. 5.1, an essential requirement to gradually introduce 3D holoscopic imaging technology into the 3DTV consumer market and to efficiently deliver 3D holoscopic content to end-users is the backward compatibility with legacy displays. This would mean that a legacy two-dimensional (2D) device (or a legacy 3D stereo device) that does not explicitly support 3D holoscopic content should be able to play a 2D (or 3D stereo) version of the 3D holoscopic content, while a more advanced device should play the 3D holoscopic content in its entirety. Hence, to enable 3D holoscopic content to be delivered and presented on legacy displays, a 3D holoscopic scalable coding approach is also desirable, where by decoding only the adequate subsets of the scalable bitstream, 2D or 3D compatible video decoders can present an appropriate version of the 3D holoscopic content.

To illustrate this concept, a display scalable architecture for 3D holoscopic video coding (proposed in [15]) with a three-layer approach is described in this chapter. As depicted in Fig. 5.10, each layer of this scalable coding architecture represents a different level of display scalability:

- *Base Layer*: The base layer represents a single 2D view, which can be used to deliver a 2D version of the 3D holoscopic content to 2D displays;
- *First Enhancement Layer*: This layer represents the necessary information to obtain an additional view (representing a stereo pair) or various views (representing multiview content). It intends to allow stereoscopic displays or multiview displays to support 3D holoscopic content;
- *Second Enhancement Layer*: This layer represents the additional data needed to support full 3D holoscopic video content.

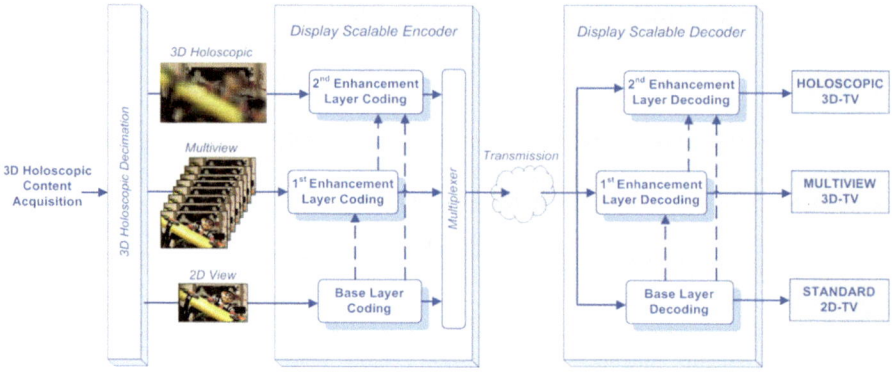

Fig. 5.10 Display scalable 3D holoscopic video coding architecture using three hierarchical layers

Another important requirement for allowing efficient storage and delivery of 3D holoscopic content to the end-users is high compression efficiency. Regarding the presented display scalable coding architecture, this would mean significant performance gains compared to independent compression of each coding layer (the simulcast case).

For this reason, the coding information flow in this 3D holoscopic coding scheme (for one access unit, i.e., for all pictures that represent the video scene at the same time instant) is defined as the following:

1. In the *Base Layer*, the 2D views are coded with a suitable standard 2D video encoder and the reconstructed frames are kept for coding the upper layers.
2. Between the *Base Layer* and the *First Enhancement Layer*, an inter-layer prediction mechanism exploits the existing inter-view correlation to improve the coding efficiency. Similarly, within the *First Enhancement Layer* an inter-view prediction is also used and the encoded and reconstructed data is fed to the *Second Enhancement Layer*. It should be noticed that, efficient prediction mechanisms between the *Base Layer* and the *First Enhancement Layer* and within the *First Enhancement Layer* are not addressed in this chapter. However, these cases have been extensively studied in the context of stereo and Multiview Video Coding (MVC) [16] extensions of the H.264/AVC [17], and the proposed 3D video coding extensions of the High Efficiency Video Coding (HEVC) [18] standard for multiview video and depth data [19]. For a good review of these 3D video coding solutions, the reader can refer to [20–22].
3. Between the *First* and the *Second Enhancement Layers*, an inter-layer prediction method, which is described in Sect. 5.4.1, is used to exploit the optical-geometric relation between the multiview content and the 3D holoscopic content. Nevertheless, it was shown in [23] that further improvements can still be reached by integrating an efficient prediction scheme that explores the inherent spatial redundancy of the 3D holoscopic content itself. For this reason, within the

Second Enhancement Layer, a self-similarity compensated prediction (described in 4.2) is integrated so as to take also full advantage of the inherent significant cross-correlation between neighboring micro-images.

Since 3D holoscopic video can be interpreted as a sequence of 2D frames (arranged as arrays of micro-images), a simple coding approach by using a regular 2D video encoder could be used. For this reason and since the presented scalable codec has most of the design elements in common with the hybrid coding techniques of HEVC [18], a 3D holoscopic scalable coding extension is easily enabled for this 2D video coding standard as explained in more details in Sect. 5.4.3.

5.4.1 Inter-Layer Prediction Scheme

It was shown in [15] that higher coding efficiency can be achieved by exploring the existing redundancy between the *First Enhancement Layer* and the *Second Enhancement Layer*, through an inter-layer prediction scheme. This prediction scheme builds an inter-layer (IL) reference picture which is then used to predict a 3D holoscopic image being coded. To build an IL reference, the following data are needed:

- *Set of 2D views*: The set of reconstructed 2D views from the previous coding layers are obtained by decoding the bitstream generated for the lower layers;
- *Acquisition parameters*: These parameters comprise information from the 3D holoscopic capturing process (such as the resolution of micro-images and the structure of the micro-lens array) and also information from the 2D view generation process (i.e., size and position of the patch, as explained in Sect. 5.3.4). This information has to be conveyed along with the bitstream to be available at the decoding side.

Therefore, two steps are distinguished when generating an IL reference: *Patch Remapping*, and *Micro-Image Refilling*.

5.4.1.1 Patch Remapping Step

This first step is an inverse process of the Basic Rendering algorithm, presented in Sect. 5.3.4, i.e., it corresponds to an inverse mapping (referred to as remapping) of the patches from all rendered and reconstructed 2D views to the 3D holoscopic image.

For this step, the needed input information is:

- One or more reconstructed 2D views;
- Patch size (PS) used to generate the 2D views;
- Micro-image resolution;
- Relative position of the patch (PP) in the micro-image used to take each different 2D view.

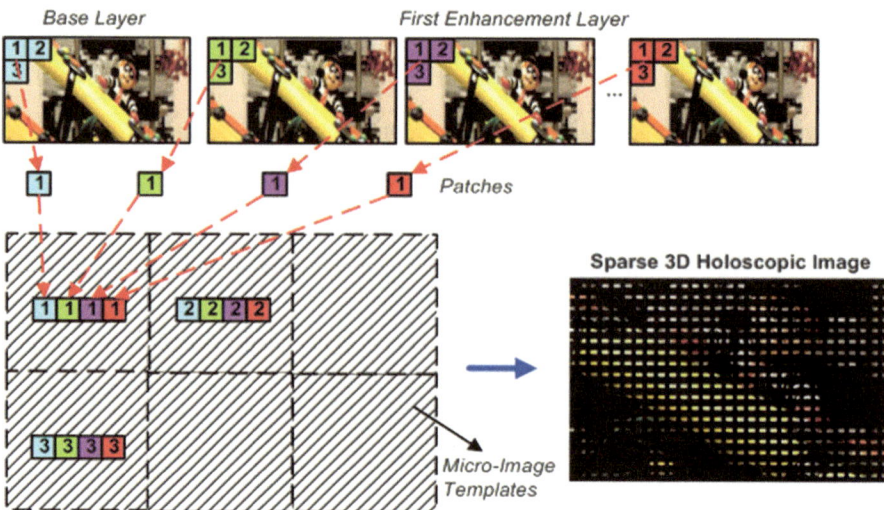

Fig. 5.11 The Patch Remapping step to generate a sparse 3D holoscopic image [15] (Copyright ©
2013, IEEE)

Then, each 2D view can be subdivided in patches and each patch can be mapped
to its original position in the 3D holoscopic image, as illustrated in Fig. 5.11.
A template of the 3D holoscopic image assembles all patches and the output of
this step is a sparse 3D holoscopic image (see Fig. 5.11).

5.4.1.2 Micro-Image Refilling Step

In the Micro-Image Refilling step, the significant cross-correlation existing between
neighboring micro-images is emulated so as to fill the holes in the sparse 3D
holoscopic image (created in the Patch Remapping step) as much as possible.

As such, the input for this step is a sparse 3D holoscopic image generated by the
Patch Remapping step. Then, for each micro-image in the sparse 3D holoscopic
image, an available set of pixels, such as the patch with size $PS \times PS$, is copied to a
suitable position in a neighboring micro-image. This position is defined by the
position of the patch being copied, shifted by $i \cdot PS$. Since this copy is done in both
horizontal and vertical directions, i is a two-dimensional variable corresponding to
the relative position of the current micro-image (where the patch is copied from) to
the neighboring micro-image (where the patch is pasted on). An illustrative example of this process is shown in Fig. 5.12 for only three neighboring micro-images.
The output of this step is the IL reference picture (Fig. 5.12).

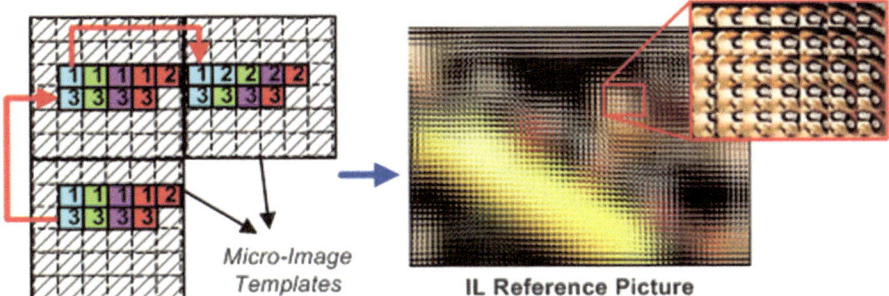

Fig. 5.12 The Micro-image Refilling step to generate a IL reference picture [15] (Copyright © 2013, IEEE)

Fig. 5.13 Example of a SS reference picture (formed by the previously coded and reconstructed area of the current frame itself)

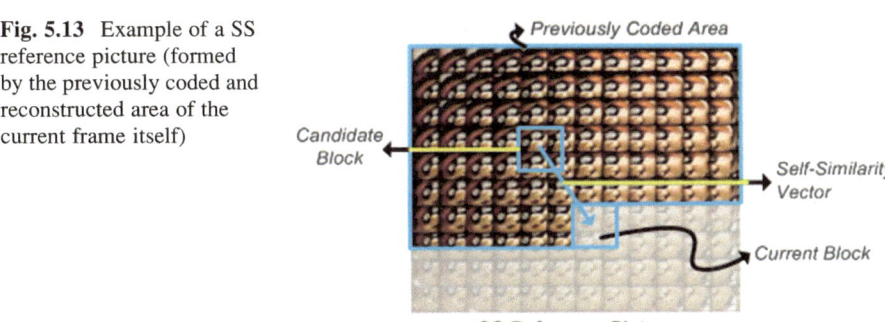

5.4.2 The Self-Similarity Compensated Prediction

A scheme for self-similarity estimation and compensation was firstly proposed in [24], in order to improve the performance of H.264/AVC [17] for 3D holoscopic video coding. In this scheme, the inherent spatial cross-correlation between neighboring micro-images is seen as type of spatial redundancy, referred to as self-similarity, which is then exploited to improve the coding efficiency. More recently, in [25], the authors proposed to introduce the self-similarity compensated prediction into HEVC so as to take advantage of the flexible partition patterns used in this video coding standard. A similar scheme, referred to as Intra Motion Compensation [26], has been proposed in the literature for screen content coding (text and graphic regions) in the HEVC range extension developments [27].

Similarly to motion estimation, the self-similarity estimation process uses block-based matching in order to find a best match—in terms of a suitable matching criterion—for prediction of a block area being encoded. However, in this case, the allowed search area is the previously coded and reconstructed area of the current picture, as illustrated in Fig. 5.13. This previously coded and reconstructed area of the current frame itself forms a new reference picture, referred to as SS reference.

As a result, the chosen block becomes the candidate predictor (see Fig. 5.13) and the relative position between the two blocks is derived as a self-similarity vector (similarly to a motion vector). In the self-similarity compensation block, the inverse quantized and inversed transformed prediction residual is added to the predictor to form the reconstructed data that is stored in the prediction memory to be available for future predictions.

5.4.3   Syntax Adaptation in a HEVC Extension for 3D Holoscopic Scalable Coding

Extending HEVC to support 3D holoscopic scalable coding basically means to introduce the IL and SS references into the buffer of reference pictures of the HEVC and to allow them to be used by the existing inter-frame prediction modes. Consequently, no changes in the lower levels of the syntax and decoding process of HEVC are needed and it only involves small changes to the high level syntax of HEVC, notably, in terms of additional high level syntax elements necessary to encode 3D holoscopic content.

The following two important pieces of information that are carried through a Sequence Parameter Set (SPS) extension in order to be available at the decoder side were identified:

- *Acquisition information*: As discussed in Sect. 5.4.1, this information is composed of a set of parameters, which are necessary to build the IL reference (e.g., resolution of micro-images, structure of the micro-lens array, size, and position of the patches).
- *Holoscopic dependency information*: Since two new references are included into the picture buffer, this information is necessary to indicate which holoscopic references (among IL and SS references) are available for each of the two reference picture lists (similarly to HEVC). This way, it is possible to distinguish the new reference frames from each other and from temporal references.

The holoscopic dependency information is used to build the reference picture lists for a 3D holoscopic picture being coded. In this case, each list is initialized as for conventional 2D video and could include any available temporal reference pictures (temporal prediction is not addressed in this chapter). Additionally, both the IL and SS references are included in the lists, becoming available for prediction of the current picture.

It is also important to notice that, in terms of lower syntax levels (e.g., Prediction Unit syntax level of the HEVC [18]), the decoding modules do not need to be aware of the reference picture type, and the distinction is only done in the higher level. Consequently, the prediction of a 3D holoscopic picture becomes adaptive in the encoding side, in the sense that the encoder can select the best reference picture in a Rate Distortion Optimization (RDO) sense, resulting in only an index of the position in the reference picture list. This index is conveyed along with the prediction information of Inter prediction modes and transmitted to identify, at the decoder side, the position of the used reference picture.

5.5 Performance Evaluation

This section illustrates the Rate Distortion (RD) performance of the 3D holoscopic coding solution described in this chapter. This performance is evaluated in two different sections:

- Section 5.1 presents a RD comparison between coding schemes based on the different representation formats of the 3D holoscopic content. As it was discussed in this chapter (Sect. 5.3), there are various possible representation formats for the 3D holoscopic content which present intrinsic correlations that are likely to be exploited to improve the coding efficiency. Therefore, a comparative analysis of the self-similarity compensated prediction method applied to different forms of representation of the 3D holoscopic content is presented. Notice that, in these coding schemes the 3D holoscopic content is encoded independently of the lower layers (*Base Layer* and *First Enhancement Layer*) in the presented scalable coding architecture. Therefore, it can be interpreted as a comparison between possible simulcast coding solutions;
- Section 5.5.2 presents the RD performance of the described scalable coding solution, where inter-layer and self-similarity compensated prediction are combined.

Two different 3D holoscopic test images with different spatial resolutions were used in these tests, as can be seen in Fig. 5.14:

- *Plane and Toy*, with a resolution of 1920×1088 pixels; micro-image resolution of 28×28 (Fig. 5.14a);
- *Laura*, with a resolution of 7240×5432 pixels; micro-image resolution of 75×75 (Fig. 5.14b; available in [28]).

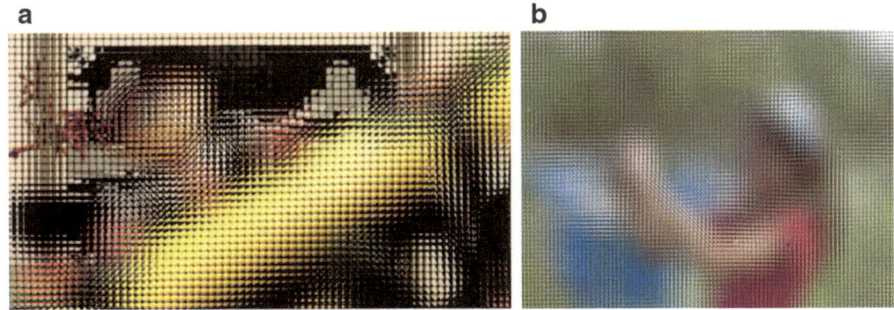

Fig. 5.14 3D holoscopic test images: (**a**) Plane and Toy; and (**b**) Laura

5.5.1 Experimental Results for Different 3D Holoscopic Representation Formats

Although the self-similarity compensated prediction is able to explore the intrinsic correlation between neighboring micro-images to improve the coding performance, it is also important to understand how the coding efficiency depends on the representation format of the 3D holoscopic content. Hence, the representation that presents better RD performance would be preferred.

As described in Sect. 5.3, besides the representation based on micro-images, it is also possible to arrange the 3D holoscopic content based on an array of viewpoint images and ray-space images. Therefore, a comparative analysis of the self-similarity compensated prediction method applied to different forms of representation of the 3D holoscopic content is presented. Moreover, solutions proposed in [10], where PVSs of micro-images, viewpoint images, and ray-space images are encoded, are also compared. However, instead of encoding with H.264/AVC (as in [10]), the HEVC codec is used, so as to have a fairer comparison with the other evaluated solutions.

In summary, seven solutions are tested and compared:

1. *HEVC*: The original 3D holoscopic image is encoded using HEVC with "Intra, main" configuration [29].
2. *3DHolo*: The original 3D holoscopic image is encoded using the self-similarity compensated prediction with "Intra, main" configuration [29]. A search range of 128 is allowed in this case.
3. *VI-based 3DHolo*: From the original 3D holoscopic image a VI-based holoscopic image is generated, which is then coded using the *3DHolo* coding scheme with "Intra, main" configuration [29]. In this case, a larger search range (312, for *Plane and Toy*, and 164, for *Laura*) is allowed, since the samples with significant correlation appear in a periodic structure given by the size of the micro-lens array (see Sect. 5.3.2). Therefore, a proportional search range is used, based on sizes of micro-lens array and micro-images. This guarantees a fairer comparison between these different representations.
4. *RI-based 3DHolo*: The original 3D holoscopic image is pre-processed so as to generate the corresponding RI-based holoscopic image, which is then coded with the presented *3DHolo* scheme. In this case, a search range of 312, for *Plane and Toy*, and 164, for *Laura*, is used.
5. *MI-based PVS*: This solution corresponds to the solution proposed in [10] and referred to as MI-based PVS. In this case, a PVS of micro-images is inter coded using HEVC. Since the resolution of each micro-image is usually considerably small, the largest coding block was set to 16×16 to avoid extra signaling [18]. The MI-based PVS is then encoded using the "Low Delay, main, P slices only" configuration [29]. Various orders for scanning the array of micro-images were tested (raster, parallel, zigzag, and spiral), but only the spiral order is presented as it achieved the best performance.

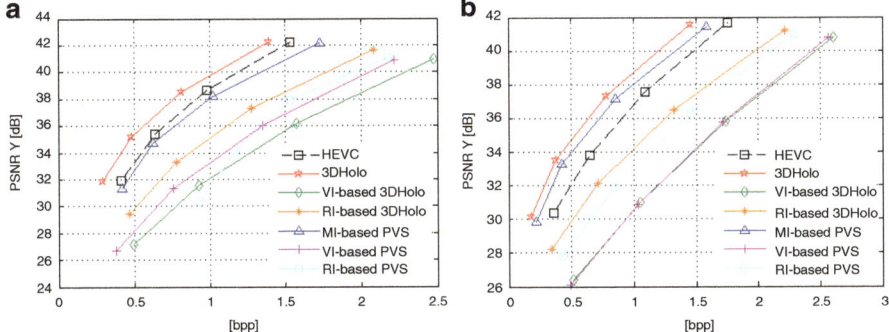

Fig. 5.15 RD performance of Simulcast Coding for: (**a**) Plane and Toy; and (**b**) Laura

6. *VI-based PVS*: This solution corresponds to the solution proposed in [10] and referred to as VI-based PVS. In this case, a PVS formed from all viewpoint images is encoded with HEVC "Low Delay, main, P slices only" configuration [29]. In this case, the largest coding block is also 16×16.

7. *RI-based PVS*: This solution corresponds to the solution proposed in [10] and referred to as RI-based PVS. In this case, a PVS formed from all ray-space images is encoded with HEVC "Low Delay, main, P slices only" configuration [29]. In this case, the largest coding block is also 16×16.

The RD performance is presented in Fig. 5.15 for each tested image in terms of luma Peak Signal to Noise Ratio (PSNR) versus average bits per pixel (bpp).

From the presented results, it is possible to conclude that solutions based on micro-image representations (i.e., *3DHolo* and *MI-based PVS* solutions) are considerably more efficient compared to the corresponding viewpoint images and ray-space images representations using the same coding scheme.

In addition, by comparing now the performance of solutions based on the MI representation, it is possible to see that *3DHolo* solution always outperforms all the other solutions, with gains that go up to 2.2 dB when compared to HEVC and 0.7 dB when compared to the MI-based PVS (for *Laura*, in Fig. 5.15b).

5.5.2 *Experimental Results for Scalable 3D Holoscopic Coding*

The RD performance of the described scalable coding solution for 3D holoscopic content is evaluated in this section.

To generate the content for the first two scalability layers, the two test images were processed with both algorithms presented in Sect. 5.3.4 (Basic Rendering and Weighted Blending). In this process, nine 2D views were generated—one for the *Base Layer* and eight for the *First Enhancement Layer*. Two sets of patch sizes were chosen for each test image: in the first set, patches of size 4×4, and 10×10, respectively, for *Plane and Toy* and *Laura* images were chosen so as to have the main object of the scene in focus (i.e., the object in the center of the image); in addition to this, the second set with sizes of, respectively, 10×10, 20×20 were also considered to evaluate the influence of patch size parameter in the performance of the proposed scalable coding scheme. For the sake of simplicity, the nine views of each test image were coded independently with the HEVC with "Intra, main" configuration [29].

Afterwards, each set of nine coded and reconstructed 2D views is processed to generate a corresponding IL reference, which is ready to be included in the reference picture buffer of the HEVC together with the SS reference.

Four different scenarios are tested and compared to evaluate the performance of the described scalable 3D holoscopic coding:

1. *HEVC Simulcast*: In this scenario, each 3D holoscopic test image is coded with the HEVC using the "Intra, main" configuration [29].
2. *3DHolo Simulcast*: Each 3D holoscopic test image is coded independently using the self-similarity compensation prediction in the HEVC, as proposed in [25];
3. *Scalable*: Each 3D holoscopic test image is coded with the scalable coding solution presented in [15], considering the two different view rendering algorithms (Basic Rendering and Weighted Blending) with the different patch sizes;
4. *3DHolo Scalable*: In this scenario, each 3D holoscopic test image is coded with the proposed combined scalable coding solution, considering the two different view rendering algorithms (Basic Rendering and Weighted Blending) with the different patch sizes.

The results are presented in Tables 5.1 and 5.2 in terms of Bjontegaard Delta (BD) measurement method [30] of PSNR and rate (BR) for each test image. Additionally, the RD performance for *Plane and Toy* is also presented in Fig. 5.16 to better illustrate the gains presented in Tables 5.1 and 5.2. Since only the performance of the *Second Enhancement Layer* is analyzed (as justified in Sect. 5.4), the results for each view rendering algorithm, as well as for each used patch size, are properly presented in different charts as it refers to different content in the first two hierarchical layers of the scalable coding architecture. Moreover, it should be noticed that the coding efficiency is analyzed here in terms of the gain

Table 5.1 BD-PSNR and BD-BR test results for Plane and Toy

	3D Holo Scalable (PSNR [dB] and BR [%])							
	Basic Rendering, patch size 4		Weighted Blend, patch size 4		Basic Rendering, patch size 10		Weighted Blend, patch size 10	
	PSNR	BR	PSNR	BR	PSNR	BR	PSNR	BR
HEVC Simulcast	1.48	−20.13	1.58	−21.16	1.68	−21.58	1.68	−21.58
3DHolo Simulcast	0.1	−1.56	0.19	−2.82	0.22	−3.31	0.22	−3.31
Scalable	1.17	−16.30	1.09	−15.33	1.25	−16.70	1.25	−16.70

Table 5.2 BD-PSNR and BD-BR test results for Laura

| | 3D Holo Scalable (PSNR [dB] and BR [%]) | | | | | | | |
| | Basic Rendering, patch size 10 | | Weighted Blend, patch size 10 | | Basic Rendering, patch size 20 | | Weighted Blend, patch size 20 | |
	PSNR	BR	PSNR	BR	PSNR	BR	PSNR	BR
HEVC Simulcast	2.26	−30.05	2.24	−29.87	2.52	−32.58	2.23	−29.73
3DHolo Simulcast	0.03	−0.51	0.01	−0.26	0.23	−4.05	0.00	−0.06
Scalable	2.03	−27.76	2.12	−28.61	1.87	−25.97	2.16	−29.07

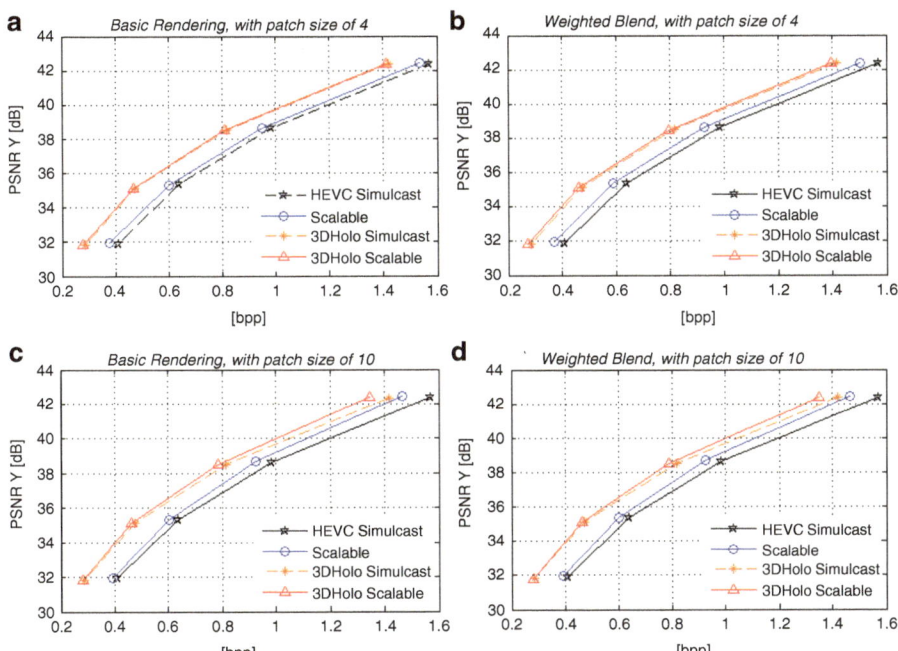

Fig. 5.16 RD performance of Scalable Codec for Plane and Toy

compared to the independent compression of the holoscopic content in relation to the content in the 2D and multiview layers (as discussed in Sect. 5.4). Thus, only the PSNR and bitrate values of the *Second Enhancement Layer* are considered in the RD charts.

Based on these results, it can be seen that the presented display scalable coding scheme, *3DHolo Scalable*, always outperforms the simulcast solution based on the HEVC (*HEVC Simulcast*). Furthermore, as can be seen in Fig. 5.16, for *Plane and Toy* test image, it is possible to take advantage of performance benefits of both prediction methods (self-similarity and inter-layer prediction) to improve the overall performance, since the *3DHolo Scalable* scheme always outperforms the scenario where each prediction method is used alone (i.e., *3DHolo Simulcast* and *Scalable*).

Moreover, it is also possible to see that, generally, by increasing the patch size the performance of *3DHolo Scalable* is improved. For example, by increasing the patch size from 4×4 to 10×10 for the test image *Plane and Toy* (see Fig. 5.16 and Table 5.1), the BD gains compared to the *HEVC Simulcast* increased from 1.48 dB/ -20.13 % (*Basic Rendering*; in Fig. 5.16a) to 1.68 dB/-21.58 % (*Basic Rendering*, in Fig. 5.16c). This can be explained by the larger amount of information from the original 3D holoscopic image that is used to generate the various lower-layers views.

5.6   Final Remarks

3D holoscopic imaging is an advantageous solution for 3D video systems, which opens new degrees of freedom in terms of content production and manipulation and also improves the users' viewing experience. However, in order to provide 3D holoscopic content with convenient visual quality in terms of resolution and 3D perception, acquisition and display devices with very high resolution are required. To deal with this large amount of data suitable representation formats and efficient coding tools become of paramount importance. In this chapter, it is shown that representations based on the micro-image format presented better rate-distortion performance for the tested images, compared to viewpoint and ray-space image representations.

Another important coding requirement that cannot be left aside is backward compatibility with legacy 2D and 3D displays. In this context, the 3D holoscopic coding solution described in this chapter provides backward compatibility by using a multi-layer display scalable coding architecture. It is shown that this display scalable coding scheme always outperforms the simulcast solution based on HEVC. Furthermore, a significant improvement in the overall performance can still be reached by combining the inter-layer and the self-similarity compensated prediction methods.

References

1. Lippmann G (1908) Épreuves réversibles donnant la sensation du relief. Journal de Physique Théorique et Appliquée 7(1):821–825
2. Aggoun A, Tsekleves E, Swash MR, Zarpalas D, Dimou A, Daras P et al (2013) Immersive 3D holoscopic video system. IEEE MultiMedia 20(1):28–37
3. Georgiev T, Yu Z, Lumsdaine A, Goma S (2013) Lytro camera technology: theory, algorithms, performance analysis. In: Proc SPIE 8667, Multimedia Content and Mobile Devices. p 86671 J–1–10
4. Raytrix Website. http://www.raytrix.de/. (2014). Accessed 7 Jul 2014

5. Lee B, Park J-H, Min S-W (2006) Three-dimensional display and information processing based on integral imaging. In: Poon DT-C (ed) Digital holography and three-dimensional display. Springer, US, pp 333–378

6. Georgiev T, Lumsdaine A. Rich image capture with plenoptic cameras. IEEE International Conference on Computational Photography (ICCP). 2010. p. 1–8.

7. Georgiev T, Lumsdaine A (2010) Focused plenoptic camera and rendering. J Electron Imaging 19(2):021106-1–021106-11

8. Zaharia R, Aggoun A, McCormick M (2002) Adaptive 3D-DCT compression algorithm for continuous parallax 3D integral imaging. Signal Process: Image Commun 17(3):231–242

9. Aggoun A (2006) A 3D DCT compression algorithm for omnidirectional integral images. In: IEEE International Conference on Acoustics, Speech and Signal Processing (ICASSP). p II–517–520

10. Olsson R (2008) Empirical rate-distortion analysis of JPEG 2000 3D and H. 264/AVC coded integral imaging based 3D-images. In: 3DTV Conference: The True Vision - Capture, Transmission and Display of 3D Video. pp 113–116

11. Yeom S, Stern A, Javidi B (2004) Compression of 3D color integral images. Opt Express 12(8):1632–1642

12. Aggoun A (2011) Compression of 3D integral images using 3D wavelet transform. J Display Technol 7(11):586–592

13. Adedoyin S, Fernando WAC, Aggoun A, Kondoz KM (2007) Motion and disparity estimation with self adapted evolutionary strategy in 3D video coding. IEEE Trans Consumer Electron 53(4):1768–1775

14. Bolles RC, Baker HH, Marimont DH (1987) Epipolar-plane image analysis: an approach to determining structure from motion. Int J Comput Vision 1(1):7–55

15. Conti C, Nunes P, Soares LD (2013) Inter-layer prediction scheme for scalable 3-D holoscopic video coding. IEEE Signal Process Lett 20(8):819–822

16. Vetro A, Wiegand T, Sullivan GJ (2011) Overview of the stereo and multiview video coding extensions of the H.264/MPEG-4 AVC standard. Proc IEEE 99(4):626–642

17. Wiegand T, Sullivan GJ, Bjontegaard G, Luthra A (2003) Overview of the H.264/AVC video coding standard. IEEE Tran Circuits Syst Video Technol 13(7):560–576

18. Han GJ, Ohm JR, Han W-J, Han W-J, Wiegand T (2012) Overview of the high efficiency video coding (HEVC) standard. IEEE Trans Circuits Syst Video Technol 22(12):1649–1668

19. Muller K, Schwarz H, Marpe D, Bartnik C, Bosse S, Brust H et al (2013) 3D High-efficiency video coding for multi-view video and depth data. IEEE Trans Image Process 22(9):3366–3378

20. White Paper on state of the art in compression and transmission of 3D video. Geneva, ISO/IEC JTC1/SC29/WG11; 2013 Jan. Report No.: N13364

21. Vetro A, Müller K (2013) Depth-based 3D video formats and coding technology. In: Dufaux F, Pesquet-Popescu B, Cagnazzo M (eds) Emerging technologies for 3D video. Wiley, West Sussex, pp 139–161

22. Müller K, Merkle P, Tech G (2013) 3D video compression. In: Zhu C, Zhao Y, Yu L, Tanimoto M (eds) 3D-TV system with depth-image-based rendering. Springer, New York, pp 223–248

23. Conti C, Nunes P, Ducla Soares L (2013) Using self-similarity compensation for improving inter-layer prediction in scalable 3D holoscopic video coding. In: Proc SPIE 8856 applications of digital image processing XXXVI. San Diego, USA, p 88561K–1–13

24. Conti C, Lino J, Nunes P, Ducla Soares L, Lobato Correia P (2011) Spatial prediction based on self-similarity compensation for 3D holoscopic image and video coding. In: 2011 18th IEEE International Conference on Image Processing (ICIP). pp 961–964

25. Conti C, Nunes P, Soares LD (2012) New HEVC prediction modes for 3D holoscopic video coding. In: 2012 19th IEEE International Conference on Image Processing (ICIP). Orlando, USA, pp 1325–1328

26. Yu S-L, Chrysafis C (2002) New intra prediction using intra-macroblock motion compensation. ISO/IEC JTC1/SC29/WG11 and ITU-T SG16 Q. 6; 2002 May. Report No.: JVT-C151

27. Budagavi M, Kwon D-K (2013) Intra motion compensation and entropy coding improvements for HEVC screen content coding. In: Picture Coding Symposium (PCS). pp 365–368
28. Geogiev T (2014) Todor Georgiev Website. http://www.tgeorgiev.net/. Accessed 7 Jul 2014
29. Bossen F (2013) Common HM test conditions and software reference configurations. Geneva; Report no.: JCTVC-L1100
30. Bjontegaard G (2013) Calculation of average PSNR differences between RD curves. Austin, TX, USA, Report no.: VCEG-M33

Chapter 6
Visual Attention Modelling in a 3D Context

Haroon Qureshi and Nicolas Tizon

Abstract This chapter provides a general framework for visual attention modelling. A combination of different state-of-the-art approaches in the field of saliency detection is described. This is done by extending various spatial domain approaches to the temporal domain. Proposed saliency detection methods (with and without using depth information) are applied on the video to detect salient regions. Finally, experimental results are shown in order to validate the saliency map quality with the eye tracking system results. This chapter also deals with the integration of visual attention models in video compression algorithms. Jointly with the eye tracking data, this use case provides a validation framework to assess the relevance of saliency extraction methods. By using our proposed saliency models for video coding purposes, we demonstrate substantial performances gains.

6.1 Introduction

With the advancement in modern digital technologies, the need for better quality assessment of 3D multimedia contents has also increased. While delivering 3D media contents to fixed and mobile users, bandwidth, equipment performance, and cost are dominant factors. These requirements can be minimized with the help of newly emerging technologies in the field of compression, intelligent video coding, and most specifically in the field of visual attention modelling. For example, one possible way is to compress the video equally without considering the contents which requires a specific amount of bandwidth. Another but a smarter way is to use an encoder that is sensitized to encode salient regions with higher data rate. This can be done by exploring visual attention modelling techniques.

H. Qureshi (✉)
Institut für Rundfunktechnik GmbH (IRT), 80939 Munich, Germany
e-mail: qureshi@irt.de

N. Tizon
VITEC Multimedia, 99 rue Pierre Sémard, 92324 Châtillon, France
e-mail: nicolas.tizon@vitecmm.com

© Springer Science+Business Media New York 2015
A. Kondoz, T. Dagiuklas (eds.), *Novel 3D Media Technologies*,
DOI 10.1007/978-1-4939-2026-6_6

Visual attention modelling is a method that tries to estimate features in a visual scene that will catch the viewer's attention. The competency of human visual system to notice visual saliency is enormously fast and dependable. Visual saliency is represented by a saliency map. Saliency map can be defined as a visual representation of a corresponding scene.

Computational models of visual attention use different image features such as color, intensity, orientation, face, etc. Those are used as a clue to firstly identify salient regions and then predict places that are likely to attract human attention. The following section presents an overview of different saliency detection techniques.

6.2 Overview of Saliency Detection Approaches

The capability of computational image processing has increased up to an extent that it has opened many new doors of research possibilities. Since many years, detecting salient regions is an important research topic in the area of visual attention modelling. It provides a fast initial pre-processing stage for many vision applications. The recent developments in the field of visual attention modelling are greatly inspired by two famous models regarding visual perception that try to mimic a human visual system model. One of the popular models is Feature Integration Theory (FIT) [35]. An extension to FIT model is Guided Search (GS) [36]. Both models attempt to explain schematically a human's behavior while looking at a visual scene. The idea of automatic saliency detection is also equivalent to these models. A very comprehensive survey, covering an analysis of scores, datasets, and a model of state-of-the-art technologies in visual attention modelling is presented in the papers [3, 6].

Generally saliency detection methods can be categorized into two parts: image based saliency and video based saliency approaches. Image based saliency methods emphasize to find salient regions from the background. While the video based saliency methods aim is to distinct salient motions from the background. Detection of salient regions can be done in many different ways:

- Detection using various clues
- Detection using spatial information
- Detection using temporal information
- Detection using depth information

6.2.1 Detection Using Various Clues

The ideas of saliency detection are commonly used in many applications such as object-of-interest segmentation [16, 24], motion detection, frame rate up-conversion [21, 22], image re-targeting [31], person tracking [10], identification

[13], video summarization [27], compression [15], and also in the area of energy saving [4].

With the development of new applications, researchers are trying hard to make the saliency detection techniques as much robust as possible. That is why in order to detect salient region, different clues for example, camera motion [1], face [7, 8, 29], and speech are used by the researchers.

6.2.2 Detection Using Spatial Information

Detecting region of interest (ROI) from the intra image information has been studied since many years. Many different approaches have proposed in the meanwhile. Koch et al. [25] proposed to extract maps for each feature and apply Winner Take All (WTA) procedure in order to highlight the most apparent area of interest. It considers bottom-up criteria due to their low level complexity. They include three features based on the difference in terms of color, intensity, and orientation. These features are the most attractive elements for the human brain. Figure 6.1 presents three images with a specific point of interest based on these features and Fig. 6.2 illustrates this approach in more detail.

Many approaches for saliency detection emerge from the field of communications engineering. Itti et al. simulated the process of human visual search in order to detect salient regions [26]. Achanta et al. and Cheng et al. estimate saliency using

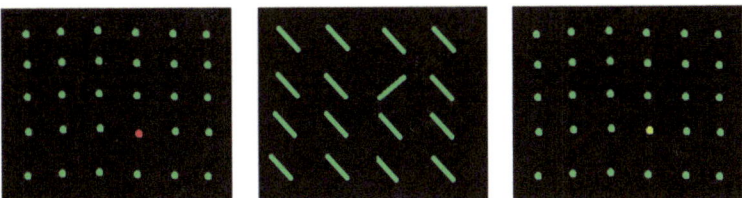

Fig. 6.1 Most attractive features for human brain. From *left* to *right*: color, orientation, and intensity

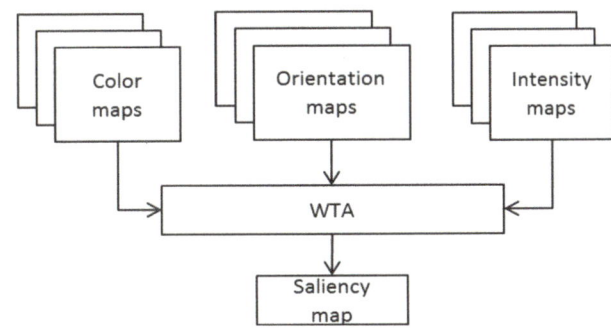

Fig. 6.2 Overview of Winner Take All (WTA) Approach

frequency-tuned saliency and contrast-based concept [2, 17]. Methods based on Fast Fourier transformation offer a fast way to process information in real time. A Spectral Residual approach (SR) [18] proposed by Huo et al. considers irregularity as a clue from the smooth spectrum for saliency detection. Spectral Residual approach (SR) depends on the observation that log spectra of different images share alike information. Huo et al. assume that the average image has a smooth spectrum and any object or area that causes an abnormality from the smooth spectrum will grab the viewer's attention. In SR approach image is first transformed into frequency domain. Then the spectrum is smoothed and subtracted from the original one. Finally the result is transformed back into the time domain. More recently, Image Signature approach (IS) [19] based on Discrete Cosine transform (DCT) and its variation [5, 29] is also explored in detecting salient regions.

6.2.3 Detection Using Temporal Information

The main aim of video based saliency is to separate salient motions from the background. In the context of video many approaches have been proposed. In fact, many similar approaches from the still image domain extended to temporal domain as well. For example, the Temporal Spectral Residual (TSR) [11] approach is an extension of Spectral Residual (SR) [18] approach.

The Temporal Spectral Residual (TSR) [11] approach explores the possibility of estimating salient regions by removing redundant information. Principally, the TSR approach applies the SR approach on temporal slices (XT and YT planes). XT and YT are the planes of image lines in a temporal domain. The final saliency map comprises motion saliency, which is considered to represent saliency with the motion information.

Another approach, the Temporal Image Signature (TIS) [29] is an extension of Image Signature (IS) approach [19]. TIS approach explores the advantage of DCT to estimate salient regions. In TIS approach instead of analyzing the Fourier spectrum as in the Temporal Spectral Residual approach, DCT information for saliency detection is used. TIS approach is discussed in more detail in the later session.

Researchers have also suggested combining both spatial and temporal information available in the scene in order to detect saliency regions. For example, Guo et al. [9] proposed to use spatio-temporal information available in the image for saliency detection. They introduce the idea of using the images phase spectrum of Fourier Transform (PFT). They compared their results with those of Huo et al's algorithm [18] for various test images and calculated the minimum, maximum, and average pixel difference. They concluded that the difference is marginal and can be neglected.

6.2.3.1 Temporal Image Signature Approach

The proposed approach, Temporal Image Signature (TIS) [29] explores the possibilities of combining Image Signature approach (IS) [19] with the Temporal
Spectral Residual (TSR) [11] in order to determine the visual saliency. In this
section, a brief and comprehensive overview of the proposed TIS approach is
described.

It is proposed to extend the IS approach to the temporal domain. It is well-known
fact that as the size of the image increases, the FFT becomes progressively complex
at a much more rapid rate. DCT is more efficient in this case and the proposed
approach explores the advantage of DCT.

TIS approach can be divided into three main steps:

1. Frame division
2. Saliency detection
3. Transformation and accumulation

Frame Division Given a video clip as an input, at first, a number of frames
(XY) are sliced into the XT and YT planes. XT and YT are the planes of image
lines in a temporal domain. By doing this there is a better representation of the
horizontal-time (XT) and vertical-time (YT) axes which also contain movement
information.

Saliency Detection Provided the frames separately in horizontal-time and vertical-
time plane, the IS [19] approach is applied separately on all planes. This identifies
saliency in movements rather than in image creation. By extracting just salient
movement the camera movement is ignored.

Transformation and Accumulation Given the saliency information in their
respective planes, finally an accumulation is done after transforming back from
XT and YT domain to XY domain. This resulted in a final saliency map which
contains regions that grasp viewer's attention. Figure 6.3 shows the proposed
system.

Saliency Map Computation Given a video clip with size $m \times n \times t$ where $m \times n$
is the image size and t is the number of frames being processed, TIS [29] approach
can be written as follows.

$$MapXT_m = sign(DCT(I_{XTm})) \tag{6.1}$$

$$MapYT_n = sign(DCT(I_{YTn})) \tag{6.2}$$

$$(MapXT_m) \xrightarrow[Transform]{} hMapXY_t \tag{6.3}$$

$$(MapYT_n) \xrightarrow[Transform]{} vMapXY_t \tag{6.4}$$

Adding Eq. (6.3) and (6.4)

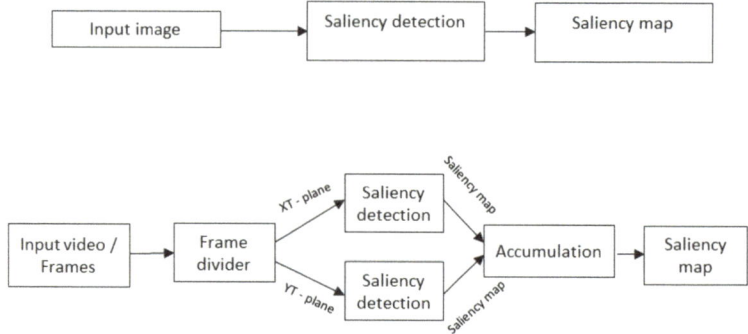

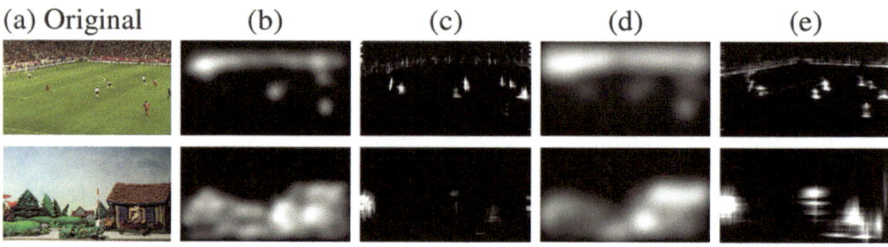

Fig. 6.3 The IS method [19], *above*, vs. the Proposed (TIS) [29] method, *below*

(a) Original	(b)	(c)	(d)	(e)

Fig. 6.4 Visual comparison of state-of-the-art approaches (**b** [20], **c** [11], **d** [19]) with the proposed TIS approach [29], (**e**). In comparison the proposed method detects and highlights more salient regions. *Note: The Foot-ball sequence is copyrighted by the EBU whereas the second sequence "Sandmännchen" is copyrighted by mdr, NDR and rbb, scopas medien AG, 2011*

$$SalMap(t) = hMapXY(t) + vMapXY(t) \qquad (6.5)$$

MapXT and MapYT represent horizontal and vertical maps, SalMap is the final saliency map, I_{XT} and I_{YT} are the slices of the images in horizontal and vertical axes. "sign (DCT(I))" is the IS method [19] applied on the image.

Conclusion and Results After TIS is applied on a video sequence, it is observed that maps for few frames in the beginning as well as in the end are quite noisy. This is due to splitting of the video information into XT and YT planes. Artifacts occur at the XT and YT planes edge when the IS method is applied on it. Therefore when the map is converted back to XY plane those artifacts end up in the first and last few frames. Noisy maps can be discarded by splitting the video into sequences of, e.g., 50 frames each overlapping each other.

Furthermore, the proposed approach delivers important contextual information, therefore can also be seen as an automatic activity detector for motions of different objects. Comparison of a saliency map obtained from the proposed TIS approach (e) with the results of saliency maps obtained from other state-of-the-art approaches is shown in Fig. 6.4.

6.2.4 Improved Temporal Image Signature Approach

The proposed approach [30] improves the TIS approach [29] by adding the face features. The proposed approach presents how a combination of the Temporal Image Signature approach (TIS) and face detection [23] will allow detecting visually prominent features separately. In this section a brief and comprehensive overview of the proposed approach is described.

Detecting salient areas within the scene is a challenging task. Researchers are proposing different ideas and clues for detecting salient regions. This detection becomes even more difficult while dealing with complex scenes. Factors such as face detection can play an important clue in guiding attention to the area of interest. The proposed approach provides useful information about the location of human face where it is present in conjunction with detecting the salient regions in the scene. This information could be a useful tool for detecting human gaze attention.

The idea of combining a face map with the saliency map is used in [8] where Cerf et al. suggest combining a saliency model with the well-known Viola and Jones matching algorithm [37] whereas Schauerte et al. [5] combine Modified Cosine Transform (MCT) based face detection [14] with their saliency model. Nevertheless in this paper, a method proposed in [23][1] is used for the proof-of-concept for face detection.

The proposed approach can be divided into the following main steps:

1. Frame division
2. Saliency detection
3. Face detection
4. Transformation and accumulation

Given a video clip as an input, frame division and saliency detection step are performed in the same way as it was done in the case of TIS approach [29].

Face Detection In parallel to the salient regions detection, the detection of face is done by applying face detection method [23] on each frame separately in XY domain. After face detection, transformation of a detected face into a face map is done using 2D Gaussian weighted function [5].

Transformation and Accumulation In the final step, provided the saliency information in their respective planes, first salient information in the corresponding planes (XT and YT) is accumulated by transformation back into the XY domain and it is then fused with the face map. This results in a final map which is a combination of saliency map and face map. Figure 6.5 shows the proposed system.

---

[1] Matlab implementation of the face detection algorithm is available on http://people.kyb. tuebingen.mpg.de/kienzle/fdlib/fdlib.htm.

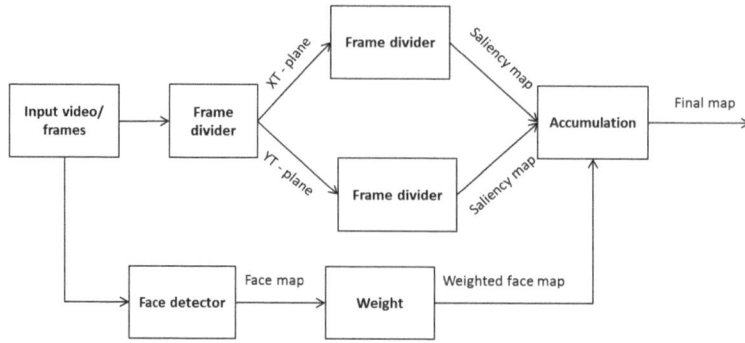

Fig. 6.5 The proposed system

Saliency Map Computation: Fusing Temporal Image Signature approach with Face Detection Given a video clip with size $m \times n \times t$ where $m \times n$ is the image size and t is the number of frames being processed, TIS can be written as follows.

Saliency map as computed in TIS approach [29] in Eq. (6.5) in the computation of saliency map (Sect. 6.2.3.1).

$$\text{SalMap}(t) = \text{hMap XY}(t) + \text{vMap XY}(t)$$

Transformation of the detected face to the face conspicuity map is done using weighted function as proposed in [5]. This can be written as

$$\text{Detected face} \xrightarrow{\textit{Transform}} fMapXY_t \qquad (6.6)$$

Weighted face map can be represented as

$$wfMapXY_t = fMapXY_t + w \qquad (6.7)$$

Saliency map can be combined with face map by adding Eqs. (6.5) and (6.7)

$$FSalMap(t) = SalMap(t) + wfMapXY_t \qquad (6.8)$$

where hMap and vMap represent horizontal and vertical maps, fMap, SalMap are the face and saliency maps and w is the assigned weight mask (gaussian or linear).

Conclusion and Results It has been shown that the proposed method is a well appropriate algorithm for not only detecting salient regions but similarly is valuable tool once combining salient information with human face concurrently in a video scene. Face definitely attracts human visual attention. Eye tracker results in Sect. 6.3.1 provide evidence of that fact. The accuracy of the proposed method depends strongly on the quality and accuracy of face detection method. It is also

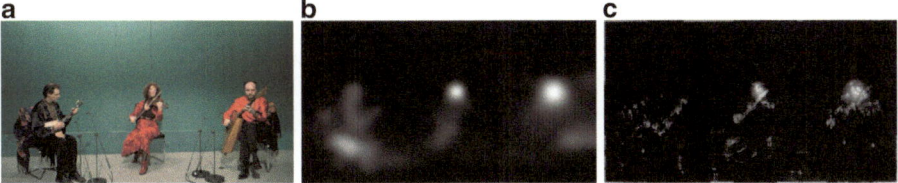

Fig. 6.6 Visual comparison of state-of-the-art approach (**b**) with the proposed approach (**c**). (**a**) Original. (**b**) GBVS [8]. (**c**) Proposed [30]

observed that a human face can be detected wrong because of a change in lightning condition or in the presence of multiple faces. This is shown in Fig. 6.6, which provides a comparison of a saliency map obtained from the proposed approach with the GBVS saliency maps [8]. However the proposed method detects and highlights more salient regions as compared to other approach.

6.2.5 Detection Using Depth Information

In the previous sections, the saliency computation methods are performed in the 2D space. In the literature as well, the main contributions to saliency extraction are focused on 2D image processing. In the context of this chapter, we are using video sequences acquired with a multi-view camera rig [34] which captures one scene through different points of view. Then, the content is rendered on auto-stereoscopic screens from which the user can virtually move around the scene. In order to do so, the multi-view renderer needs disparity maps between the different cameras. These disparity maps computed just after the acquisition stage are transmitted jointly with the camera views and allow the render to extrapolate none existing intermediary views. Therefore, it means that in such a context, the disparity information is available with the corresponding views and provides useful additional information to compute saliency maps. Especially, during the video encoding process, the encoders could exploit this information in order to improve the compression performances.

In the literature, a few studies deal with the contribution of stereo disparity or depth to saliency map extraction. In [28], a simple approach which combines nearness and motion is proposed. In [32], the proposed approach is based on 3D motion estimation and the more advanced and complete approaches described in [12, 39] are combinations of image saliency, motions saliency, and depth saliency. In [33], a bit rate adaptation algorithm is described where the video encoder allocates more or less bit rate following the interest or saliency level of the region. The content used in this study comes from a video game for which a perfect depth map is available. Based on this depth information, a simple and fast algorithm divides the image into regions with different saliency levels and these levels are then mapped to quantization steps. For our approach, we will transpose this work in the multi-view context by using disparity maps in order to differentiate the image

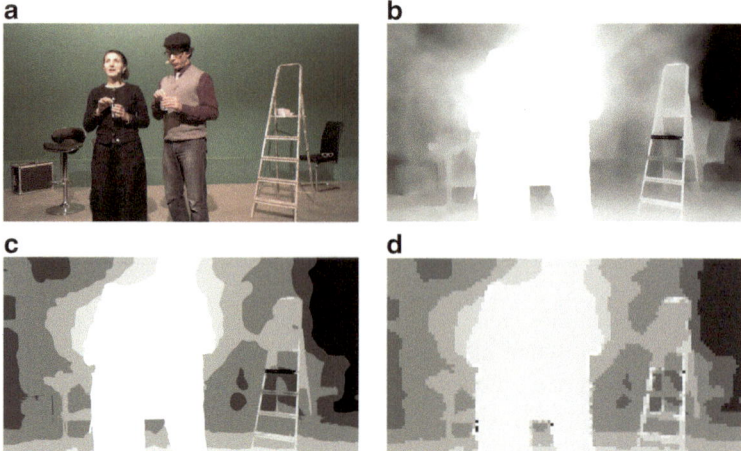

Fig. 6.7 Disparity map processing steps (partitioning: 7 regions), "Actors" sequence. (**a**) YUV frame. (**b**) Disparity map. (**c**) Quantized disparity map. (**d**) Rescaled disparity map

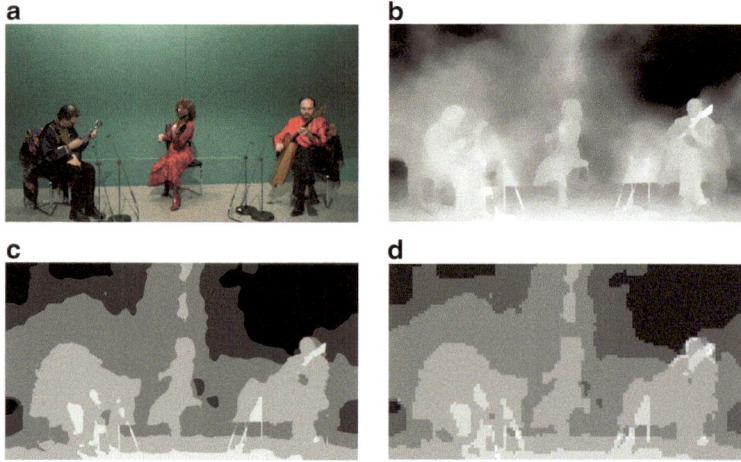

Fig. 6.8 Disparity map processing steps (partitioning: 6 regions), "Musicians" sequence. (**a**) YUV frame. (**b**) Diparity map. (**c**) Quantized disparity map. (**d**) Rescaled disparity map

regions. In many practical cases, the objects that belong to the scene's foreground are subject to higher visual attention from the user and are more relevant in terms of QoE. From this assumption, we can consider that the multi-view disparity map contains the main information when defining regions of interest. In our case, the disparity maps are extracted during the post-acquisition step and transmitted with the corresponding view to the encoder. Then, to improve the compression we will process this disparity map images in order to obtain a macroblock based partitioning of each frame as described in Figs. 6.7 and 6.8.

First of all, the disparity values are quantized (uniform quantization) over a small number of integer values (6 values in Fig. 6.8c). The integer value of each pixel, from 0 to 5 in the given examples, represents its depth level. The lower values (black) correspond to the background and the higher values (white) to the foreground. As this ROI partitioning is used next to perform the bit rate adaptation during the video encoding stage, the next processing step consists in downscaling the quantized disparity maps (d). This downscaling is simply obtained by averaging the disparity values for each macroblock (16×16 pixels) and adding the number of different disparity values in the macroblock. Let's denote $\mathbf{MB}_k$ the 16×16 matrix which represents the kth macroblock of the quantized disparity map (Figs. 6.7c and 6.8c). Let's denote E_k an integer subset of minimal size which contains all $\mathbf{MB}_k$ element values. Hence in Figs. 6.7d and 6.8d, $\mathbf{MB}_k$ is downscaled to a scalar $disp_k$ (one pixel) obtained as follows:

$$disp_k = \frac{1}{256}\sum_{i=0}^{16}\sum_{j=0}^{16}\mathbf{MB}_k[i,j] + card(E_k) \tag{6.9}$$

After this last processing, each macroblock is represented by an integer value $(0..N)$ which is supposed to indicate its degree of importance. This last operation on the disparity map allows a better discrimination of homogeneous areas from regions with high frequencies (object borders). For instance in Fig. 6.7, the stepladder is almost in the background but thanks to the second term in Eq. (6.9), these pixels will be better taken into account during the video encoding process (see Sect. 6.3.3).

6.3 Evaluation

6.3.1 Eye Tracking Setup and Validation of VAM Model

For the purpose of evaluation an eye tracking test was conducted. The goal was to produce reference material so that saliency results of the developed algorithms [29, 30] can be compared with the eye tracker data. Moreover some additional purposes of eye tracking system were to get a better understanding of human vision and see if there are significant differences between 2D and 3D.

6.3.1.1 Eye Tracking Hardware Setup

For the eye tracking test a Tobii TX300 Eye Tracker[2] system was used. This system is a stand-alone eye tracking system which is placed in front of the viewer. As

---

[2] http://www.tobii.com/en/eye-tracking-research/global/products/hardware/tobii-tx300-eye-tracker.

Fig. 6.9 Eye tracking setup
(side view, simplified)

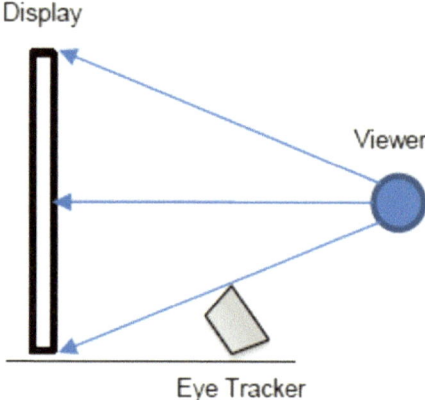

viewing monitor an LG 47 LM660S-ZA[3], 47 Zoll LCD television was used. For displaying stereoscopic content the TV makes use of passive polarized glasses. For content play-out two DVS Clipster[4] systems were used in synchronized mode.

6.3.1.2   Setup of Eye Tracking Test

The test setup was built based on the position of the TV. The viewer was placed within usual viewing distance in front of the display panel. The distance was slightly varying depending on the viewer's size and eventually was adjusted based on the eye tracker calibration results. Figure 6.9 illustrates the used setup.

6.3.1.3   Video Content Used for the Eye Tracking Test

Different clips composed of different content properties (e.g., different genre types, sports, documentation, and also still scenes) were used and were combined into one video. The video was rendered in 2D and 3D.

6.3.1.4   Execution of the Eye Tracking Test

The content was presented in 3D as well as in 2D in different viewing sessions. Twenty five participants were asked to watch the content without any specific instructions with the exception to follow the calibration points in the calibration pattern presented as first clip in the beginning of the video presentation. Recalibration for each participant was performed again if required after the

---

[3] http://www.lg.com/de/service-produkt/lg-47LM660S.

[4] http://www.dvs.de/de/produkte/video-systems/clipster.htm.

calibration check. For every participant two captures were generated, one from the 3D and one from the 2D video presentations.

6.3.1.5   Post-Processing of the Captured Gaze Data

For further usage some post-processing was applied on the captured data. Specific points in the recorded data, e.g. start of content, were marked to enable matching the captured samples to the video frames. Therefore the eye-tracking results were exported from the eye-tracking system and filtered for only relevant information. Afterwards the samples were matched with the video frames based on the captured timestamps and the previously marked video positions like start of content.

6.3.2   Validation of Temporal Image Signature Approach with Eye Tracking Data

In order to evaluate the proposed method TIS approach [29], two sequences as shown in Fig. 6.10 are used for eye tracking experiments. The heat maps (which show the salient regions where the users fixated their gazes more often) obtained by means of an eye tracker system[5] is fused with the proposed saliency model as shown in Fig. 6.10.

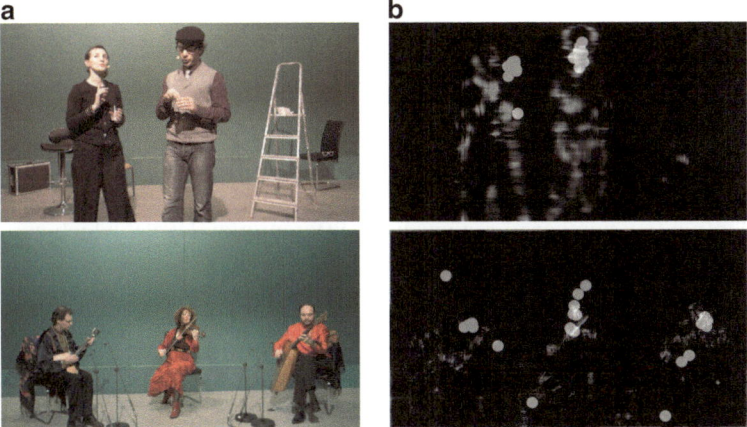

Fig. 6.10 Eye tracking heat maps (binary) fused with TIS approach [29]. (**a**) Original (**b**) Eye tracking data fuse with TIS approach [29]

---

[5] Tobii Technology TX300 eye tracker system was used for obtaining the heat maps.

Fig. 6.11 2D (*left*) and 3D (*right*) heatmap

6.3.2.1 Conclusion

It was observed that heat maps obtained from the eye tracker almost matches with the salient areas detected by the TIS approach [29]. It is also observed that face plays an important role and it indeed grabs visual attention wherever it is present as also shown in Fig. 6.10 that reveals multiple users points of focus with the strong emphasis on face. Additionally the captured data (2D and 3D sequences) using eye tracker were compared. No big differences in distribution of gaze fixation positions were found here so in general it can be assumed that the viewing behavior does not differ much. There were some minor hints though where it might be assumed that in specific situations or properties of the clips a viewer might tend to watch the foreground more intensive in 3D than in 2D as shown in Fig. 6.11.

6.3.3 VAM Based Differentiated Bit Allocation for Video Coding

In a general context and without specific requirements, a saliency map is classically provided as a dense gray level image (8 bits per pixel). As described in Sect. 6.2.5, for some post-processing purposes (e.g., video coding), the saliency regions can be merged into a few number of regions with different gray levels, which correspond to the level of interest of the region. Especially for video compression, this image partitioning is very useful in order to optimize the bit rate allocation by decreasing the bit rate budget for less salient regions.

In a lossy video coding process, the quantization factor (Qp) is the parameter that allows to allocate more or less bit rate, regarding a target quality level. Especially, in H.264/AVC [38], the quantization parameter is set frame by frame and specific offsets can be applied at the macroblock (16 × 16 pixels) granularity. In this context, a dense disparity map is no longer necessary and the map is downscaled before being passed to the encoder (see Figs. 6.7d and 6.8d). Once this quantized and downscaled disparity map is obtained, the saliency levels are mapped to the corresponding offsets. In order to improve efficiently the encoding process, the Qp offset mapping must be done by smoothing the regions where saliency variations

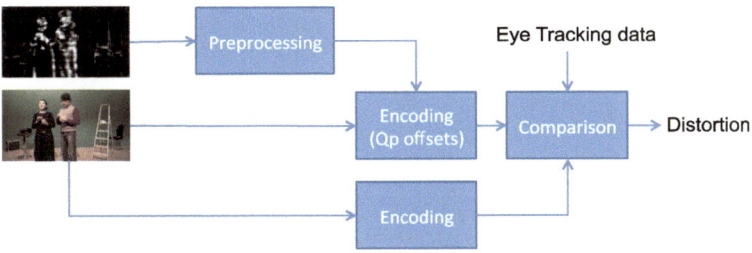

Fig. 6.12 Validation framework

are too high. In the same way, the quantization step increase between adjacent regions must be minimized in order to achieve more consistent and efficient encoding. Different saliency to quantization offset mappings are considered and presented in the next sections.

By using the eye tracking information, the performances of the proposed approach are measured and compared to the reference. The validation framework is illustrated in Fig. 6.12. In this framework, the pre-processing module consists in mapping the image regions with the corresponding quantization factor offsets. After encoding, the video quality is objectively measured after masking the image with the eye tracked areas. In this way, the measured quality corresponds to the quality of the pixels subject to higher visual attention as measured by the eye tracking system. In the next sections, the performances measured with the two sequences "Actor" and "Musicians" are presented.

6.3.3.1   Disparity Map Approach

The algorithm described in Sect. 6.2.5 aims at converting disparity maps into saliency based image partitioning where the most salient region corresponds to the white pixels. In this approach, the saliency computation is based on a simple heuristic (objects in the foreground are more salient than the background) and having in mind that the saliency maps is used further to manage the bit rate allocation at encoding stage. This second constraint leads to increase the saliency of the pixels that belong to object borders. Finally, the Qp offsets maps introduced previously are obtained by producing the negative image of the saliency maps as depicted in Fig. 6.13.

The rate-distortion curves obtained with the two sequences after H.264 encoding[6] are provided in Fig. 6.14. In this figure, we can see the interest of the approach, especially when the bit rate increases. For instance, for the Actor sequence at a quality level of 42 dB, the encoding with VAM allows saving around 25 % of the bit rate compared to the classical encoding approach, when measuring

---

[6] http://www.videolan.org/developers/x264.html.

Dense saliency map Qp offsets per macroblock

Fig. 6.13 Saliency to quantization offset mapping with the disparity based approach (*Top*: "Musicians" sequence, *Bottom*: "Actors" sequence)

Fig. 6.14 Video coding performances with the disparity based approach (*Top*: "Musicians" sequence, *Bottom*: "Actors" sequence)

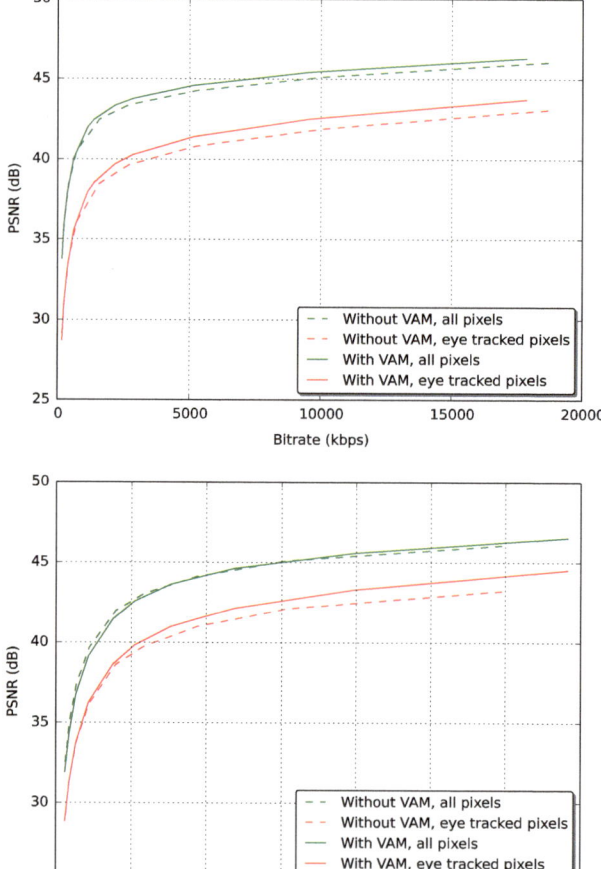

the quality on the more salient pixels. This bit rate saving is obtained by decreasing the quality of the less salient regions (the background) that are subject to lower perceptual attention. However, this quality decrease remains limited and the PSNR values measured on the entire images are very close with and without VAM. For the "Musicians" sequence, the VAM approach achieves better performances even by considering all the pixels for PSNR measurements. In this case, the VAM approach improves the overall behavior of the encoder and provides more efficient information to the rate-distortion control algorithm than the classical approach.

6.3.3.2 TIS Approach

With the temporal image signature (TIS) approach described in Sect. 6.2.3.1, the salient regions are sparse and the approach tends to highlight high frequency regions like object borders. This behavior is particularly visible on the "Actors" sequence (see Fig. 6.15, bottom-left). However, for a block based video encoding like H.264/AVC the quantization step cannot vary too much from a pixel block to another due to the macroblock based (16×16 pixels) encoding and in order to limit the blocking artifacts. Thus, to obtain the Qp offsets per macroblock, we apply several post-processings:

- Median filtering (blurring) of the saliency map,
- Downscaling of the dense saliency map (each dimension divided by 16),
- Nonuniform quantization which aims at merging the more salient regions,
- Negative transformation to obtain Qp offsets.

Dense saliency map   Qp offsets per macroblock

Fig. 6.15 Saliency to quantization offset mapping with the TIS approach (*Top*: "Musicians" sequence, *Bottom*: "Actors" sequence)

After downscaling, the quantized saliency map $\mathbf{S}_q$ is obtained as follows: $\mathbf{S}_q = \lfloor \mathbf{S}^{1/3} \rfloor$, where $\mathbf{S}$ represents the down-scaled (non-quantized) saliency map. The non-quantized saliency maps are gray level images coded with 255 levels. Hence, the maximum Qp offset value is $\lfloor 255^{1/3} \rfloor = 6$.

In Fig. 6.15, two examples of Qp offsets images obtained from the dense saliency map are showed. For the two sequences "Musicians" and "Actors," the effect of the post-processing is noticeable. The salient regions cover more pixels than in the saliency map due to the merging effect and the transitions between regions are more progressive.

The rate-distortion curves obtained with the two sequences after H.264 encoding are provided in Fig. 6.16. In this figure, we can see the interest of the approach, especially when the bit rate increases. For instance, for the "Actor" sequence at a quality level of 42 dB, the encoding with VAM allows saving more than 40 % of the

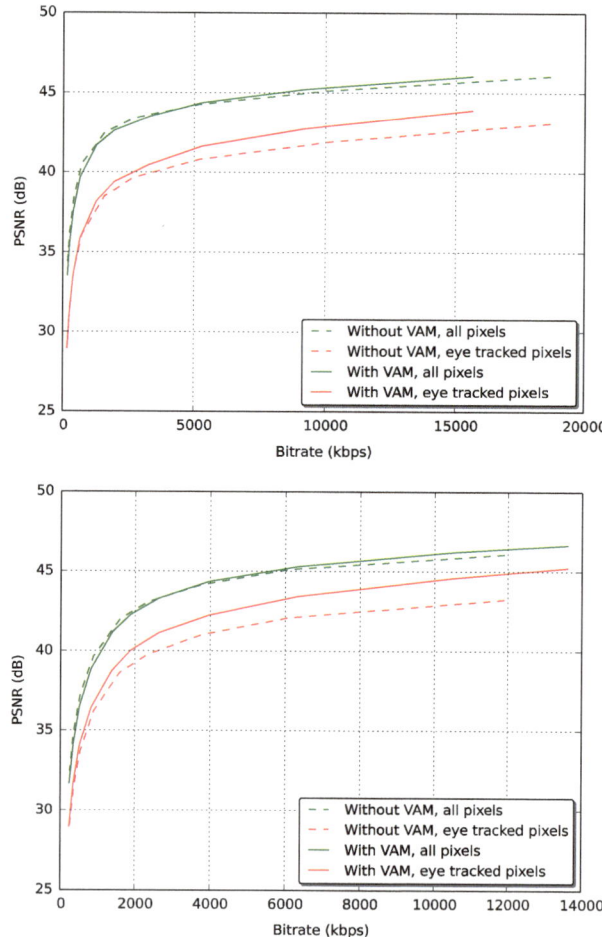

Fig. 6.16 Video coding performances with the TIS approach (*Top*: "Musicians" sequence, *Bottom*: "Actors" sequence)

bit rate compared to the classical encoding approach. As for the disparity based approach, this bit rate saving is obtained by decreasing the quality of the less salient regions which are subject to lower perceptual attention. However, this quality decrease remains limited and can be adjusted following the application requirements. In order to adjust the level of differentiation between regions, the quantization is a key process. For instance, computing the quantized maps as follows: $S_q = \lfloor S^{1/4} \rfloor$ will merge more dramatically the salient regions and decrease the maximum Qp offset from 6 to 4. In terms of performances, this would reduce the quality gain on the salient pixels and better preserve the less salient areas.

Conclusion

In this chapter, we have firstly listed the main approaches, classically used to extract saliency information from 2D video signals. The more efficient approaches consist in processing jointly different clues of the signal: spatial and temporal information. In addition, the content considered in this chapter is multi-view and third dimensional data are freely available in order to build a quick and efficient saliency computation algorithm based on disparity maps. In addition to this fast and basic implementation, an improved version of the temporal image signature based approach is proposed.

In order to assess the proposed saliency map extraction methods, an eye tracking system has been used to produce reference material and to compare the algorithms outputs with the eye tracker data. Statistically, we have observed a good matching between the saliency maps produced by the TIS method and the eye tracking data.

Moreover, a specific use case where saliency is used to improve bit rate allocation in a video encoder has been tested. With the disparity based approach, we can easily, without requiring high computation resources, improve the rate-distortion performances when focusing on pixels referenced by the eye tracking tests. With the improved TIS approach, the performances gains on the salient pixels is even better and especially at high bit rates, the approach achieves quite impressive results.

Finally, this study confirms the interest and benefit of visual attention modelling when designing advanced image processing applications (e.g., 3D multi-view video) and the integration of these approaches in complete application frameworks is really promising when considering QoE optimization aspects.

Acknowledgements This work was supported by the ROMEO project (grant number: 287896), which was funded by the EC FP7 ICT collaborative research program.

References

1. Abdollahian G, Edward JD (2007) Finding regions of interest in home videos based on camera motion. In: IEEE International Conference on Image Processing (ICIP), vol 4, 2007
2. Achanta R, Hemami SS, Estrada FJ, Süsstrunk S (2009) Frequency-tuned salient region detection. In: CVPR, pp 1597–1604, 2009
3. Ali B, Laurent I (2013) State-of-the-art in visual attention modeling. IEEE Trans Pattern Anal Mach Intell 35(1):185–207
4. Ariizumi R, Kaneda S, Haga H (2008) Energy saving of TV by face detection. In: Proceedings of the 1st international conference on pervasive technologies related to assistive environments, pp 95:1–95:8, 2008
5. Boris S, Rainer S (2012) In: Predicting human gaze using quaternion DCT image signature saliency and face detection. In: Proceedings of the IEEE workshop on the applications of computer vision (WACV), 2012
6. Borji A, Tavakoli H, Sihite D, Itti L (2013) Analysis of scores, datasets, and models in visual saliency prediction. In Proceedings of the IEEE international conference on computer vision, pp 921–928, 2013
7. Cerf M, Frady EP, Koch C (2008) Predicting human gaze using low-level saliency combined with face detection. In: Advances in neural information processing systems, vol 20, 2008
8. Cerf M, Frady EP, Koch C (2009) Faces and text attract gaze independent of the task: Experimental data and computer model. J Vis 9(12):1–15
9. Chenlei G, Qi M, Liming Z (2008) Spatio-temporal Saliency detection using phase spectrum of quaternion fourier transform. In: CVPR'08, 2008
10. Corvee E, Bremond F (2009) BioID: a multimodal biometric identification system. In: 3rd international conference on crime detection and prevention (ICDP 2009), 255(5):1–6, 2009
11. Cui X, Liu Q, Metaxas D (2009) Temporal spectral residual: fast motion saliency detection. In: Proceedings of the 17th ACM international conference on multimedia (MM'09), pp 617–620, 2009
12. Dittrich T, Kopf S, Schaber P, Guthier B, Effelsberg W (2013) Saliency detection for stereoscopic video. In: Proceedings of the 4th ACM multimedia systems conference, ser. MMSys 13, pp 12–23. ACM, New York
13. Frischholz RW, Dieckmann U (2000) Features and objects in visual processing. Computer 33 (2):64–68
14. Froba B, Ernst A (2004) Face detection with the modified census transform. In: Sixth IEEE International Conference on Automatic Face and Gesture Recognition, pp 91–96 (2004)
15. Hadizadeh H, Bajić IV (2014) Saliency-aware video compression. IEEE Trans Image Process 23(1):19–33
16. Han J, Ngan KN, Li M, Zhang HJ (2006) Unsupervised extraction of visual attention objects in color images. IEEE Trans Circuits Syst Video Technol 16(1):141–145
17. Heng M-M, Zhang G-X, Mitra NJ, Huang X, Hu S-M (2011) Global contrast based salient region detection. In CVPR, pp 409–416, 2011
18. Hou X, Zhang L (2007) Saliency detection: a spectral residual approach. In: Conference on computer vision and patten recognition (CVPR), pp 1–8, IEEE, 2007
19. Hou X, Harel J, Koch C (2012) Image signature: highlighting sparse salient regions. IEEE Trans Pattern Anal Mach Intell 34(1):194–201
20. Itti L, Koch C, Niebur E (1998) A model of saliency-based visual attention for rapid scene analysis. IEEE Trans Pattern Anal Mach Intell 20(11):1254–1259
21. Jacobson N, Lee Y-L, Mahadevan V, Vasconcelos N, Nguyen TQ (2010) A novel approach to FRUC using discriminant saliency and frame segmentation. Sci. Am 19(11):2924–2934.
22. Jacobson N, Nguyen TQ (2011) Video processing with scale-aware saliency: application to Frame Rate Up-Conversion. In: ICASSP, pp 1313–1316, 2011
23. Kienzle W, Bakir GH, Franz M, Schölkopf B (2004) Face detection-efficient and rank deficient. In: NIPS, 2004

24. Ko BC, Nam J-Y (2006) Object-of-interest image segmentation based on human attention and semantic region clustering. J. Opt. Soc. Am. A 23(10):2462–2470
25. Koch C, Ullman S (1985) Shifts in selective visual attention: towards the underlying neural circuitry. Human neurobiology. In California Institute of Technology. 2000. Ph.D. Thesis, 4(4), 219–227, 1985
26. Laurent I (2000) Models of bottom-up and top-down visual attention. PhD thesis, California Institute of Technology
27. Ma Y-F, Hua X-S, Lu L, Zhang H-J (2005) A generic framework of user attention model and its application in video summarization. IEEE Trans Multimed 7(5):907–919
28. Maki A, Nordlund P, Eklundh J-O (1996) A computational model of depth-based attention. In: Proceedings of the 13th international conference on pattern recognition, vol 4, pp 734–739, 1996
29. Qureshi H (2013) DCT based Temporal image signature approach. In: 8th international conference on computer vision theory and applications (VISAPP'13), pp 208–212, Barcelona, 2013
30. Qureshi H, Ludwig M (2013) Improving temporal image signature approach by adding face conspicuity map. In: Proceedings of the 2nd ROMEO workshop, Istanbul, 9 July 2013
31. Radhakrishna A, Sabine S (2009) Saliency detection for content-aware image resizing. In: IEEE international conference on image processing, 2009
32. Riche N, Mancas M, Gosselin B, Dutoit T (2011) 3d saliency for abnormal motion selection: the role of the depth map. In: Proceedings of the 8th international conference on computer vision systems, ser. ICVS11, pp 143–152. Springer, Berlin
33. Tizon N, Moreno C, Preda M (2011) Roi based video streaming for 3d remote rendering. In: MMSP, pp 1–6, IEEE, 2011
34. Tizon N, Dosso G, Ekmekcioglu E (2014) Multi-view acquisition and advanced depth map processing techniques. In: Kondoz, A., Dagiuklas, T. (eds) 3D Future Internet Media, pp. 55–78. Springer, New York
35. Treisman AM, Gelade G (1980) A feature-integration theory of attention. Cogn Psychol 12:97–136
36. Treisman A (1986) Features and objects in visual processing. Sci Am 255(5):114–125
37. Viola P, Jones M (2001) Rapid object detection using a boosted cascade of simple features. In Computer Vision and Pattern Recognition, (CVPR), vol 1, pp 1–511, 2001
38. Wiegand T, Sullivan G, Bjontegaard G, Luthra A (2003) Overview of the h.264/avc video coding standard. IEEE Trans Circ Syst Video Technol 13(7):560–576
39. Zhang Y, Jiang G, Yu M, Chen K (2010) Stereoscopic visual attention model for 3d video. In: Proceedings of the 16th international conference on advances in multimedia modeling, ser. MMM10, pp 314–324. Springer, Berlin

Chapter 7
Dynamic Cloud Resource Migration for Efficient 3D Video Processing in Mobile Computing Environments

Constandinos X. Mavromoustakis, Paraskevi Mousicou, Katerina Papanikolaou, George Mastorakis, Athina Bourdena, and Evangelos Pallis

Abstract This chapter presents a dynamic cloud computing scheme for efficient resource migration and 3D media content processing in mobile computing environments. It elaborates on location and capacity issues to offload resources from mobile devices due to their processing limitations, towards efficiently manipulating 3D video content. The proposed scheme adopts a rack-based approach that enables cooperative migration for redundant resources to be offloaded towards facilitating 3D media content manipulation. The rack-based approach significantly reduces crash failures that lead all servers to become unavailable within a rack and enables mobile devices with limited processing capabilities to reproduce multimedia services at an acceptable level of quality of experience (QoE). The presented scheme is thoroughly evaluated through simulation tests, where the resource migration policy was used in the context of cloud rack failures for delay-bounded resource availability of mobile users.

7.1 Introduction

Utility computing [1] offers synchronous and asynchronous availability of services, in the context of cloud computing paradigm. The services offered through utility-based cloud computing platforms, mainly emerged from the appearance of public

C.X. Mavromoustakis (✉) • P. Mousicou
Computer Science Department, University of Nicosia, 46 Makedonitissas Avenue, P.O. Box 24005, 1700 Nicosia, Cyprus
e-mail: mavromoustakis.c@unic.ac.cy; mousicou.P3@unic.ac.cy

K. Papanikolaou
European University Cyprus, 6 Diogenous Str., 1516 Nicosia, Cyprus
e-mail: K.Papanikolaou@euc.ac.cy

G. Mastorakis • A. Bourdena • E. Pallis
Department of Informatics Engineering, Technological Educational Institute of Crete, Heraklion, Crete, Greece
e-mail: gmastorakis@staff.teicrete.gr; abourdena@pasiphae.eu; pallis@pasiphae.eu

© Springer Science+Business Media New York 2015
A. Kondoz, T. Dagiuklas (eds.), *Novel 3D Media Technologies*,
DOI 10.1007/978-1-4939-2026-6_7

119

computing utilities. These services vary in the context of physical location, as well as the distribution nature of requested data and resources. In this framework, a cloud computing system can be classified as public, private, community-based, or so-called social, hybrid or a combination of all the above, based on the model of deployment [2, 3]. On the other hand, the lack of processing power and capacity resources for mobile devices to efficiently manipulate 3D video content aggravate the reliability for executing these resources in the mobile terminal and cause capacity-oriented failures. Execution failures due to the lack of processing resources should be adjusted through the offloading process and the resource/task migration mechanism. In this context, a mechanism has to be investigated for ensuring that the capacity of supplied resources can be expanded to meet the needs of mobile devices for the efficient manipulation of 3D media content.

Based on the lack of resources for running and executing concurrent processes associated with the efficient 3D video content manipulation over mobile devices, this chapter elaborates on the resource migration according to a proposed cloud-computing scheme. This scheme guarantees that the resource utilization, the operational efficiency, the portability of claimed resources, and the on-the-move fulfillment of mobile users claimed tasks, in terms of memory for an efficient 3D video content manipulation, are maximized. This is achievable through the dynamic task migration, which enables partitionable 3D content associated tasks to be a-priori migrated, according to the scheduling policy of the resource sharing process, in order to avoid potential memory failures on a wireless device. In this framework, a partitionable parallel processing cloud-rack system is studied, where partitions are handled by a subsystem cloud rack that receives and handles the migration process of the claimed resources.

The structure flow of this chapter is as follows: Sect. 7.2 describes related work regarding cloud-based 3D media content, as well as highlights the need in adopting a dynamic task migration scheduling policy based on capacity-awareness for mobile computing environments. Section 7.3 presents the proposed migration scheduling scheme that enables a cooperative mobile resources migration for efficient 3D media content manipulation fulfilling the requested resources for the mobile devices. Section 7.4 presents the results obtained, by conducting simulation experiments for the performance evaluation focusing on the behavioral characteristics of the scheme along with the system response. Finally, Section 7.5 concludes this chapter, elaborating on research findings, as well as on potential future directions.

7.2 Related Work and Research Motivation

The recent trend in multimedia content delivery through mobile computing platforms is related with the adoption of procedures that efficiently manage the content information. The main idea of multimedia content delivery is to obtain multiple streams of content, transferred to various user-terminals by exploiting a number of communication and network technologies. However, the provision of sophisticated applications through advanced multimedia systems arises a number of issues that have to be investigated. Among the envisioned challenges are the signal processing, power and networking management issues, while optimization of content delivery is also crucial when considering such demanding application. Moreover, the exploitation of cloud resource management techniques has to be considered for efficiently overcoming problems regarding systems resources limitations, such as processor power, memory capacity, and network bandwidth. Resource management systems are responsible to allocate cloud resources to users and applications with flexibility, maintaining service isolation [4, 5]. Such systems are expected to operate under the predefined quality of service (QoS) requirements as set by the users. For this purpose, resource management at cloud scale requires a rich set of resource and task management schemes that are capable to efficiently manage the provision of QoS requirements, while maintaining total system efficiency. The greatest challenge for this optimization problem is that of the scalability in the context of performance evaluation and measurement. Within this context, the use of dynamic resource allocation policies improves application execution performance, as well as the utilization of resources [6].

Towards overcoming the abovementioned challenges a number of research approaches have been proposed. More specifically, authors in [7] propose a media-cloud network architecture that administers in a distributed and parallel manner the units of multiple devices to obtain a high QoS in multimedia services. This approach elaborates on storage issues of multimedia content, the central processing unit that selects the speed of transfer, as well as the editing and the graphics processing unit, which is responsible for the depiction of multimedia content to provide high QoS to the users. This architecture enables the elimination of software installation in users' devices, while at the same time the storage and the processing of user multimedia application are performed in the cloud. However, the challenge in mobile media applications and services is the provision of QoS, considering delay, jitter, and bandwidth performance. Moreover, authors in [8] present a framework that promises ubiquitous multimedia services provision over mobile devices. These services have the advantage of resilience in cloud computing and in cloud storage, while they do not meet issues, such as the limited device capabilities or content availability. On the other hand, these services are hampered by issues of next generation devices [9]. More specifically, the smart phones manage a large amount of users' contextual information, but most of this information is not correlated with media context to enrich the multimedia services on mobile computing platforms. In addition, the interoperability of contextual information has to be investigated for different mobile devices with the existing resources on the web, as well as the chance to exploit context-aware services in application for mobile devices. In that case, the exploitation of QoE techniques from the users' side has the possibility to provide the proper manipulation and management, regarding the quality provision in the cloud-based network architecture.

Other research approaches related to the performance of dynamic resource allocation policies had led to the development of a computing framework [10], which considers the countable and measureable parameters that will affect task allocation. The idea of dynamically switching between local resources and remote resources for serving availability and direct access to the requested resources (often referred as multi-foraging behavior) has shed light on many research works [11–13]. Authors in [11] address this problem, by using the CloneCloud approach [14] of a smart and efficient architecture for the seamless use of ambient computation to augment mobile device applications, off-loading the right portion of their execution onto device clones operating in a computational cloud. Authors in [14] statically partition service tasks and resources between client and server portions, whereas in a later stage the service is reassembled on the mobile device. The spine of the proposal in [14] is based on a cloud-augmented execution, using a cloned VM image as a powerful virtual device. This approach has many vulnerabilities as it has to take into consideration the resources of each cloud rack, depending on the expected workload and execution conditions (CPU speed, network performance). Similar work, conducted on suspend-migrate-resume mechanisms [15] for application-specific task migration, takes into account processing diversities of the application but not the thread processing characteristics, thus presenting forecasting obstacles in safely determining the overall application execution and service time. Authors in [16] address the resource poverty issue showing the obstacles for many applications that typically require processing resources; whereas authors address at the same time hardware capabilities for introducing resource hungry applications, which are in need of ample computing resources such as face recognition, speech recognition, and language translation on the move. In particular, the application task and offloading process consist of other aspects, in terms of performance and quality. In case where the resources on data centers or location-oriented cloud racks are not enough to serve a certain resource sharing procedure resulting in failures, there are schemes that aim to face the associated resource failures in the context of data centers and are well addressed in recent literature [17, 18]. However, these schemes do not consider the cloud-to-cloud (in the context of data center to data center) and the cloud-to-device resource migration, which plays an important role in the resource manipulation and offloading procedure, in case where cloud resources and cloud service requirements are not met. Authors in [17] provide a manageable solution, according to the failure rates of servers in a large-scales datacenter, whereas at the same time authors attempt to classify them, using a range of criteria. However, these criteria along with the criteria set in [19–21] do not include servers' communications diversities in the communication process with mobile users' claims, as well as the utilization of memory and capacity of each cloud terminal in the rack.

Towards considering the cloud context-aware requirements that presented above, another related challenge concerns the real-time 3D video rendering in mobile devices, where the challenge here is the need for large amount of network and computing requirements (i.e., bandwidth, CPU). To address this issue, authors in [22] propose a framework that supports remote rendering to 3D video streams, in order to provide over the mobile devices the rendered content. In another point of view, authors in [23] specified the difficulties of large loads in the channels, the limitations in computation, and the battery life on mobile phones about free viewpoint video (FVV)/free viewpoint TV (FTV). To avoid these drawbacks, a cloud-based FVV/FTV rendering framework is proposed towards obtaining the maximum QoE for users. A prototype resource allocation scheme is described that correlates both render allocation based on end user QoE and rate allocation, as a matter of channel rate, of texture, and depth quality. Results of the proposed framework indicate the optimization of the video quality in standard mode of mobile devices. Moreover, the research work in [24] considers a novel cloud framework that allows scalable real-time 3D applications. This infrastructure exploits a scalable-pipelined process that comprised of a virtualization server network for running 3D virtual appliances, a graphics rendering network for processing graphics workload with load balancing and a media streaming network for transcoding rendered frames into H.264/MPEG-4 media.

In this context, this chapter is making progress beyond the current state-of-the-art, by proposing a dynamic resource migration mechanism based on multi-migrate-resume mechanisms, for resource/task migration of mobile users, and efficient 3D content manipulation through the cloud-based network due to processing limitations of the mobile devices. It uses a similar—but optimized—methodology with the migration in [14], whereas it allows native resources that are claimed by a mobile device to be offloaded both on a datacenter rack and to a mobile device as a file/resource sharing datacenter. This chapter also elaborates on the design and development of a modular scenario, where the proposed migration scheduling policy algorithm for establishing partitionable resource/process execution, enables task execution reliability and guarantees the efficiency in the execution of mobile users' tasks/applications. The proposed scheme avoids any failures when the resources, for running a task on a mobile device, are not sufficient.

7.3 Dynamic Cloud resource Migration Based on Temporal and Capacity-Aware Policy

Computing-as-service is becoming the dominant paradigm in today's computing environments. Considering the delays caused by WAN processing mechanisms and the potential lack of executing resources onto mobile devices, there is a great need to encompass in the resource/task execution processes, the cloud resource migration services. This service can encounter resource failures for reliable execution and can take into consideration network-and-service oriented metrics such as bandwidth and ping delays as well as migration/processing delays and manageability metrics of the resources claimed by cloud. Although bandwidth and access will continue to improve over time, the service and "transfer to device-back" latency is unlikely to get improved. Therefore, the main need in such a case is the deployment of a

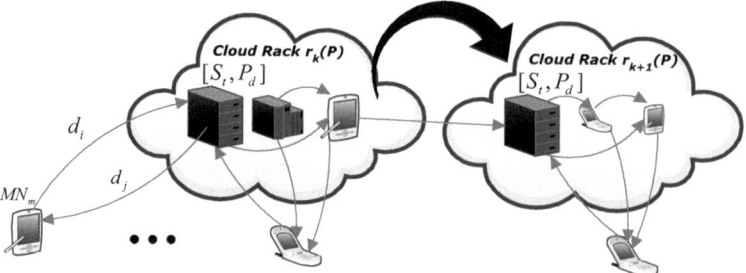

Fig. 7.1 Cloud rack configuration: Mobile and static nodes are allocated on cloud rack and resources are inter-exchanged among mobile and static users based on the best effort allocation

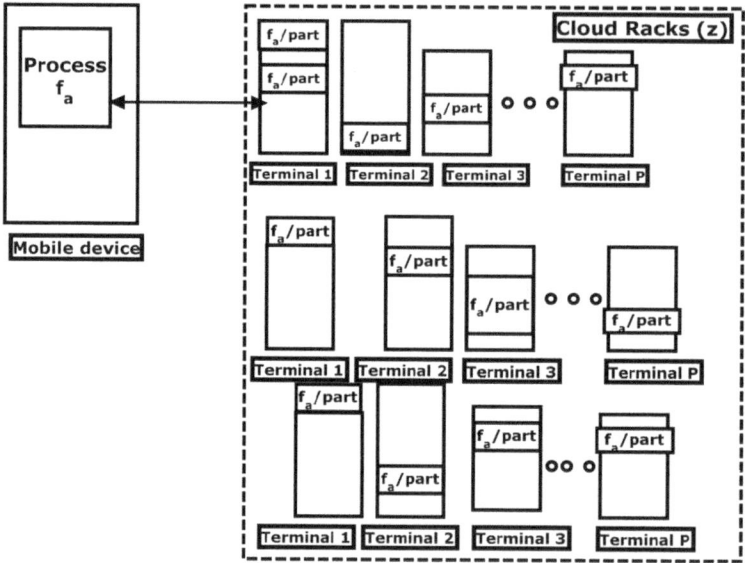

Fig. 7.2 Multi-migrate-resume scheduling cloud policy based on "a-priori" task partitioning

resource migration and offloading approach that aims to release an aggravated terminal in data center rack, and migrate according to certain parameterized requirements the resources to other racks based on the minimized service and processing delay, respectively. This work takes into account the impact of different files capacities, by using multi-migrate-resume scheduling policies under various conditions. The aim of this study is to achieve high performance, while preserving fairness in terms of sequential task execution-access (claims). Tasks consist of sequential files that are uploaded and run on terminals on the cloud rack (Fig. 7.1). Tasks start to execute, if there are terminals on cloud rack that are available to handle them. When a claim is large enough and the terminals of the cloud rack can no longer handle the request of a mobile user, the multi-migrate-resume policy (Fig. 7.2) is used. This policy enables the claims to be partitioned "a-priori" to

Table 7.1 Parameters used in the utilized scenario

d_i	Transmission (device) delay
d_j	Reception (device) delay
$r_k(P)$	Rack of P static and mobile devices
r_i	Rack redundant resources that a mobile device can claim
a_j	Claimed resource/partitionable task
s^j_i	Number of servers in rack r_k hosting application resource/task a_j
S_{aj}	The number of servers hosting application resource/task
S_t	Service time for completing a task
P_d	Processing delay of S_{aj}
MN_m	Mobile node claiming resources
s	Sequential partitions

another terminal, according to the measures of the available resources and the minimized S_t and P_d (see Table 7.1).

Moreover, this work enables the rack utility maximization (RUM), which aims at maximizing the utilization of the rack in order to be able to host processes sent by mobile devices. The RUM is based on the index of remaining r_i resources that each MN_m can claim, and on the utilization degree of each cloud rack. Considering the above parameters, in order to maximize the available resources on the rack, the following should be satisfied:

$$\Lambda_{LUM} : max\sum\nolimits_{r \in R} U_r(X_r) \forall A_x \leq C, x \geq 0 \qquad (7.1)$$

where r denotes the index of the MN_m source node, χ_r represents the transmission rate of $MN_m(r)$, C is a vector containing the associated capacities of all links of $MN_m(r)$, and $U_r(\chi_r)$ is the utility of node r when transmitting at rate χ_r. In order to avoid mobile nodes' resources execution failure, the resources (that are to be run on a node) are being partitioned and distributed to K_{MN_i} terminals in a r_k, whereas the cloud scheduler is acting as a distributor of the file chunks on r_k. The distribution policy (shown in Fig. 7.2) is based on the utilization of the K_{MN_m} on r_k, and the scheduled resources with the time frame t. The scheduling policy is shown in Fig. 7.3, where if the resources (memory and capacity limitations) are not met and the task-partitioned pieces are missing, then the cloud scheduler re-assigns the claimed chunks to other mobile terminals satisfying the delay requirements. The delay that the transmission experiences δ_{ij} should satisfy the $\delta_{ij} < d_p$, where d_p is the maximum delay in the end-to-end path from a source to a destination and is evaluated as:

$$d_p = \sum_{i=0}^{i-1} \delta_i + T_i \quad \forall max \sum_{r \in R} U_r(\chi_r) \qquad (7.2)$$

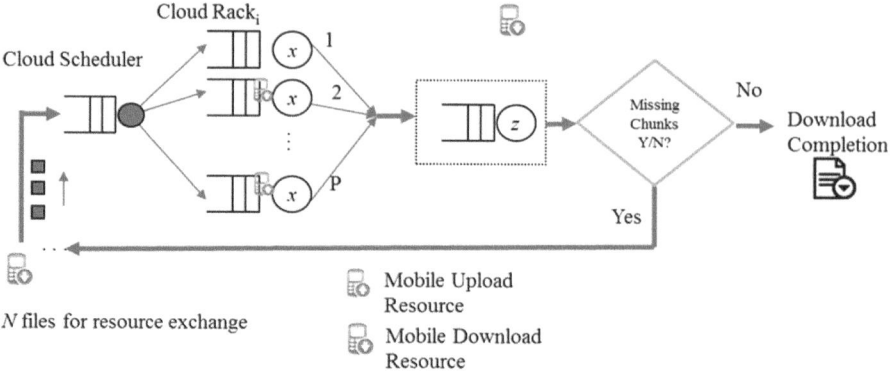

Fig. 7.3 Resource partitioning and scheduling model

In Fig. 7.3, a closed queuing processing cloud model is considered that consists of P parallel heterogeneous processor-enabled devices and the I/O subsystem. All heterogeneous processors share a single queue (memory), whereas the multi-migrate-resume policy is used for switching incomplete claims, to another cloud terminal—if the available resources of a terminal cannot complete the requests. The effects of the memory requirements and the communication latencies are explicitly represented in the closed queuing-processing cloud model. Moreover, these evaluations appear at the job/claim execution time. The z subsystem in Fig. 7.3, which is hosted on the MNi consists of mechanism that reassembles all the file chunks that were portioned, into a single module, modeled as a single I/O node with a given mean service time based on the characteristic of the device.

To aid clarity, this work considers that every resource may be an executable file x, which consists of t_x partitioning tasks where $1 \leq t_x \leq z*P$, where z is the number of different racks. Therefore, the number of tasks per file is limited to the number of rack terminals in the system. An executable resource or—as called—task can be shared and partitioned to $x_1, x_2, \ldots x_n$ and can be simultaneously processed with s sequential partitions, where $0 \leq s < z*P$, if and only if the following relation holds:

$$s + \sum_{i=1}^{n} p(x_i) \leq z * P \tag{7.3}$$

where $p(x)$ represents the number of cloud terminals that are needed to host the a_j. The scheduling strategy that was used is based on the largest first served/shortest sequential resource first served and the service notations of [25] with a-priori knowledge of the $[S_t, P_d]$ service durations.

From the rack failure perspective, this work elaborates on the examination of the impact of a possible overload in the case of rack failure/crash. When a rack failure occurs, it will directly affect several applications based on dependability and the characteristics of the rack synthesis. The resource allocation will take place in order

Table 7.2 Dynamic resource migration algorithm

1: **Inputs:** MN_m, Location($MN_m \forall [S_t, P_d]$), uploaded resources S_{aj}
2: find the r_k that hosts F_a
3: **for** all a_j claims
4:   find an optimal C_{r_k,a_j} such that $C_{r_k,a_j} \cong 1$
5: Estimate the $[S_t, P_d]$ for each r_k, C_{r_k,a_j} of the claim of MN_m
6:     **if** r_k has not enough resources satisfying the $MN_m \forall [S_t, P_d]$
7:       then
         use the partitionable notation of $_a$ nearby rack satisfying
8:       $$s + \sum_{i=1}^{n} p(x_i) \leq z * P, \text{ for each } r_k, C_{r_k,a_j}$$
9:   **Assign** a_j to the r_k terminals
10: end **for**

to respond to the performance requirements [12]. A significant measure in the system is the capacity of the rack that was not utilized in order to service an application resource/task, enabling a reallocation decision on that proportion. The capacity metric is used to measure the capacity loss, as in Eq. (7.4), where r_k is the rack, a_j is an application and s_k^j is the number of terminals P in rack $r_k(P)$ hosting application a_j, and $S_{a_j}(k)$ is the number of servers hosting application across all different racks $r_k(P)$.

$$C_{r_k,a_j} = \frac{s_k^j}{\sum_k S_{a_j}(k)} \qquad (7.4)$$

Equation (7.4) shows that if there is minimal loss in the capacity utilization, i.e., $C_{r_k,a_j} \cong 1$ then the sequence of racks $S_{a_j}(k)$ is optimally utilized. The latter is shown through the conducted simulation experiments in the next section. The dynamic resource migration algorithm is shown in Table 7.2 with the basic steps for obtaining an efficient execution for a partitionable resource that cannot be handled by the existing cloud rack and therefore the migration policy is used to ensure that it will be continuing the execution. The continuation is based on the migrated policy of the partitionable processes that are split, in order to be handled by other cloud rack terminals and thus omit any potential failures.

7.4 Performance Evaluation Analysis, Experimental Results, and Discussion

The performance evaluation encompasses the extracted results conducted by simulation experiments, using the network model—as per the mentioned scenario above. The scenario was simulated using discrete event simulation model using

Table 7.3 Cloud rack terminals characteristics

Device #	CPU (GHz)	RAM (GB)	Core no.	Hard disk (GB)	Cache (MB)	Core speed (GHz)	Upload speed (Mbits/sec)
1	2.1	8	Intel Duo	600	2	5	0.6
2	2.3	16	Quad 6600	500	2	5	0.6
3	2.1	4	i5	400	2	3	0.6
4	4	16	i5	1,000	2	5	0.6
5	2.1	8	i7	600	2	3	0.6
6	2.3	16	i5	500	2	5	0.6
7	2.1	4	Quad 6600	400	2	3	0.6
8	4	16	i5	1,000	2	5	0.6

Java-based simulator. The simulated scenario was evaluated for the achievable throughput, reliability degree, and delay-bounded transmissions (task that has delay limitations). The mobility model used by the mobile users is based on probabilistic Fractional Brownian Motion (FBM) [26], where nodes are moving, according to certain probabilities, location, and time. Topology of a "grid" based network was modeled, according to the grid approach described in [27]. Each MN_m node directly communicates with other nodes, if the area situated is in the same $(3 \times 3$ centre) rectangular area of the node. For the simulation of the proposed scenario, the varying parameters described in previous section were used, exploiting a two-dimensional network, consisting of maximum 100 MN_m/per r_k, (i.e., terminal nodes) located in measured area.

Different sets of experiments were conducted, in order to examine the impact of the different capacities in the proposed resource migration scheme, as well as measures for the sizes of the tasks to be migrated. Table 7.3 contains the different cloud terminals characteristics, regarding the terminal processing power and memory capabilities, in order to serve as migrated terminals that will host the partitioned resource up-on the represented cloud rack. Large files are files that are between 20 and 200 Mbytes, whereas small files are files with capacity between the range of 0.1 and 20 Mbytes. The mean service time with the number of racks is also shown in Fig. 7.4. The mean service time is greater for large files that are not migrated in partitionable parts to other terminals on the cloud racks.

The percentage of migrated resources with the number of cloud racks that are utilized is shown in Fig. 7.5. It is obvious that large task/application files are migrated, in order to be served in an offloaded manner. Concurrent claimed resources with the number of "in-service" cloud racks are shown in Fig. 7.5. This figure shows the load of the resources that can be concurrently served through a cloud rack with the technical characteristics of the terminals shown in Table 7.3 (see Fig. 7.6). In addition, the Total Service Time in contrast with the technical characteristics of the particular devices utilized (shown in Table 7.3) is shown in Fig. 7.7. Moreover, Fig. 7.7 presents the complementary cumulative distribution

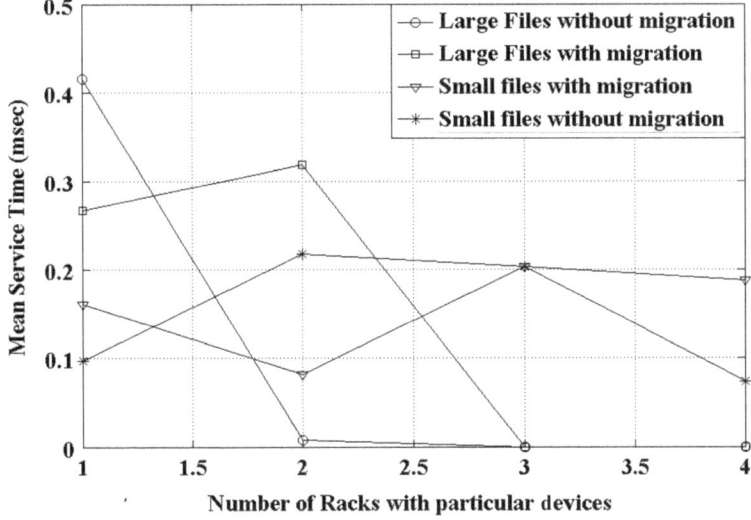

Fig. 7.4 Mean service time with the number of racks

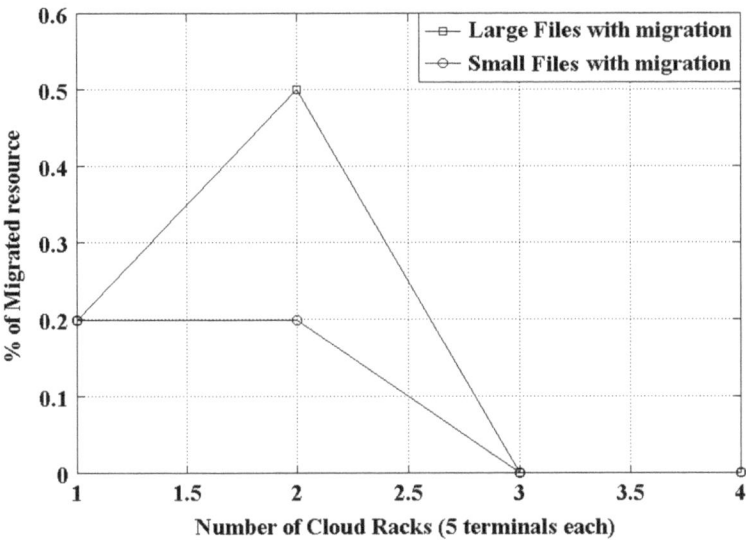

Fig. 7.5 Percentage of migrated resources with the corresponding number of cloud racks

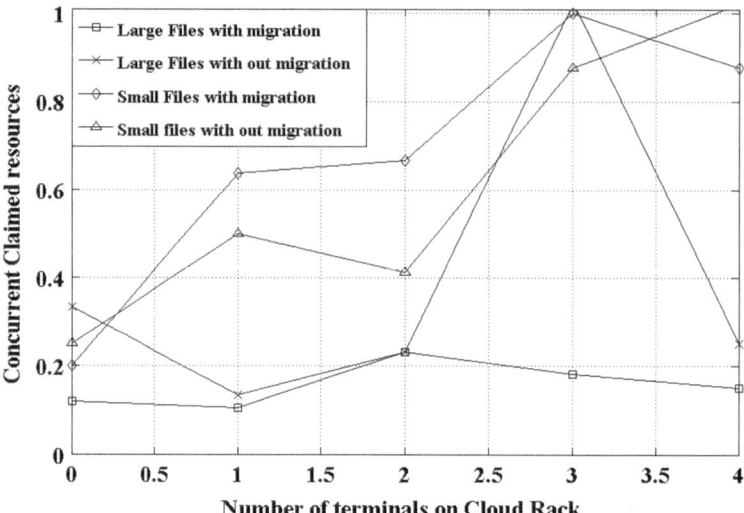

Fig. 7.6 Migrated resources and concurrent claimed resources with the number of "in-service" cloud racks

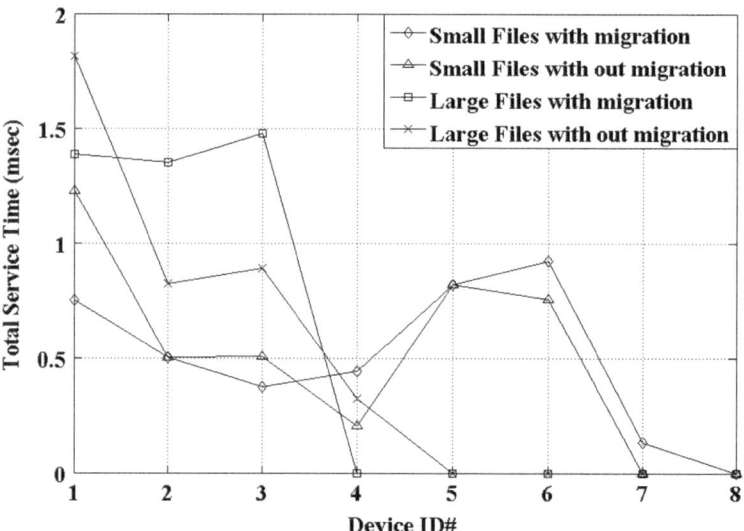

Fig. 7.7 Total service time in contrast to the devices' characteristics (Table 7.3)

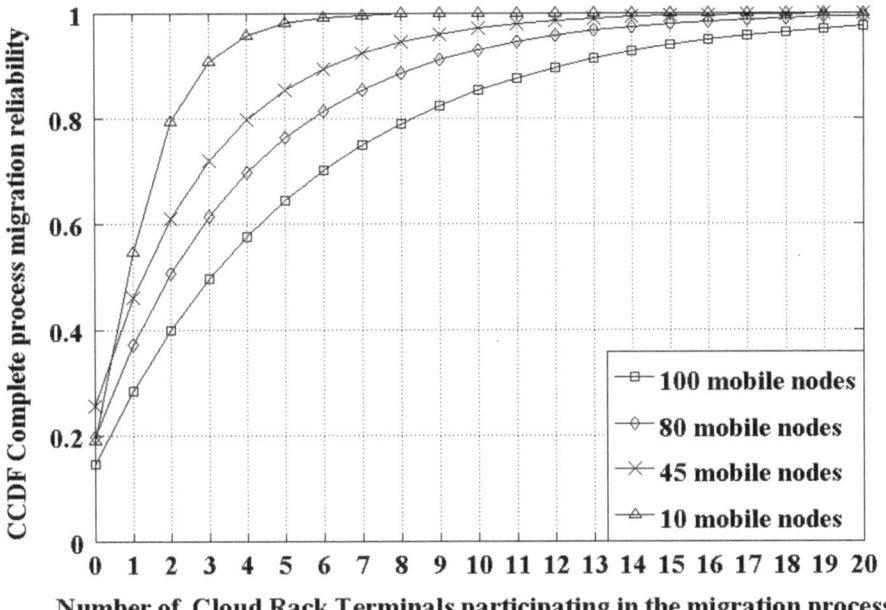

Fig. 7.8 Total service time in contrast with the particular device

function (CCDF) for the complete resource migration reliability, with the number of cloud rack terminals that are participating in the process for different number of mobile nodes. It depicts that when the number of mobile nodes increases, the CCDF migration process reliability decreases progressively. This occurs as the mobile nodes claim more resources from cloud terminals and, in turn cloud terminals forward these resources to other terminals. This process causes additional delays to be considered as well as the completion of the number of serviced processes to decrease progressively as the number of nodes increase.

When considering partitionable tasks, migration is of primary interest to examine the Throughput response of the system. Throughput response of the system with the number of "in-service" cloud racks is shown in Fig. 7.7. Figure 7.7 depicts that for large files and when the terminals have no remaining memory resources, throughput dramatically drops. In turn, the comparative execution time with the different capacities for the tasks/resources needed to be serviced, for mobile devices and CloneCloud [11] through Wi-Fi, is depicted in Fig. 7.8. The proposed migration scheme is shown to be robust in terms of execution time. This is proved by the extracted sets of simulation experiments for the different capacities of the requests of the mobile nodes (see Figs. 7.9 and 7.10).

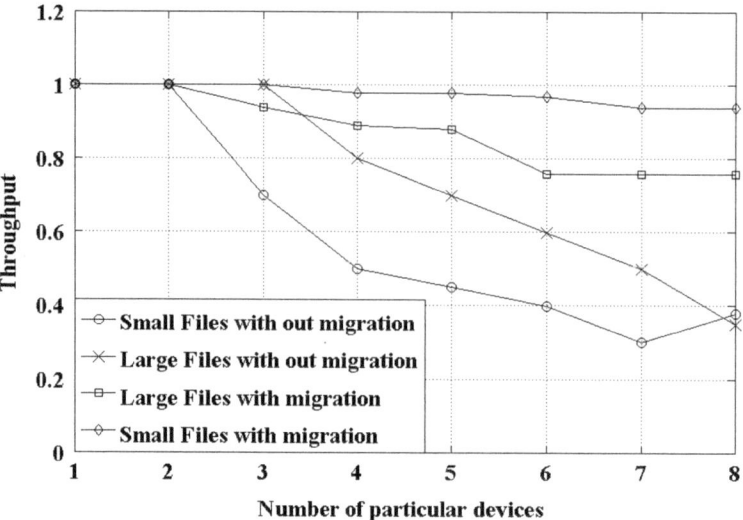

Fig. 7.9 Throughput response with the number of "in-service" cloud racks

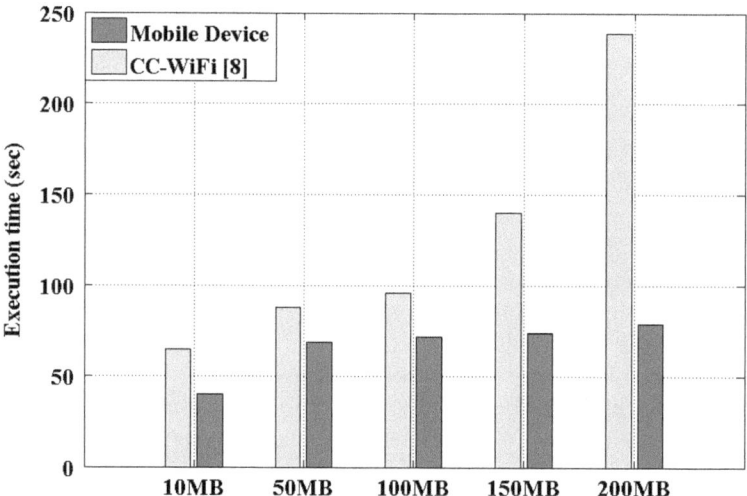

Fig. 7.10 Execution time with the different capacities for resources, for mobile devices and CloneCloud [8] through wifi

Conclusion

This chapter proposes a novel task migration scheme, where partitionable resources can be migrated, in order to be fulfilled and executed according to their limited service time and the allowed execution rates. The migration scheme provides an efficient cloud rack terminals resources exploitation and reliable task execution provided to the mobile end-recipients. The proposed methodology hosts the RUM scheme, which considers the availability of resources and the rack utilization degree. The proposed scheme offers high successful execution delivery rates during the resource migration process. Towards evaluating the performance of the proposed methodology, a number of experimental tests were conducted under controlled simulation conditions. The obtained experimental results verified the efficiency of proposed task migration scheme using the capacity-aware metrics. In this respect, fields for future research directions include the expansion of this model into a green motion-aware dynamic resource migration policy, using selective time-limited offloading policy according to capacity-aware diversities of the system.

Acknowledgment We would like to thank the European FP7 ICT COST Action IC1105, 3D-ConTourNet-3D Content Creation, Coding and Transmission over Future Media Networks (WG-3), for the active support and cooperation.

References

1. Yeo C-S, Buyya R, Dias de Assuncao M, Yu J, Sulistio A, Venugopal S, Placek M (2008) Utility computing on global grids. Published at the Handbook of Computer Networks, Wiley, Hoboken
2. Mell P, Grance T (2009) The NIST definition of cloud computing, national institute of standards and technology, information technology laboratory. Tech-nical Report Version 15
3. Benson T, Akella A, Maltz D-A (2010) Network traffic characteristics of data centers in the wild (IMC). Paper presented at the IMC 2010
4. Drepper U (2008) Cost of the virtualization, ACM Queue
5. Kotsovinos E (2011) Virtualization: blessing or curse? ACM Queue
6. Slegers J, Mitriani I, Thomas N (2009) Evaluating the optimal server allocation policy for clusters with on/off sources. J Perform Eval 66(8):453–467
7. Wenwu Z, Luo C, Wang J, Li S (2011) Multimedia cloud computing. IEEE Signal Process Mag 28(3):59–69
8. Dey S (2012) Cloud mobile media: opportunities, challenges, and directions. Paper presented at the IEEE International Conference on Computing, Networking and Communications. pp 929–933
9. Dejan K, Klamma R (2009). Context-aware mobile multimedia services in the cloud. Paper presented at the 10th International Workshop of the Multimedia Metadata Community on Semantic Multimedia Database Technologies

10. Warneke D, Kao O (2009) Nephele: efficient parallel data processing in the cloud. Paper presented at the 2nd Workshop Many-Task Computing on Grids and Supercomputers, Nov. 14–20, 2009, ACM, Portland, OR, USA, pp 1–10. ISBN: 978-1-60558-714-1. doi: 10.1145/1646468.1646476

11. Chun B, Ihm S, Maniatis P, Naik M, Patti A (2011) Clonecloud: elastic execution between mobile device and cloud. Paper presented at the Sixth Conference on Computer systems of EuroSys, pp 301–314

12. Papanikolaou K, Mavromoustakis C (2013) Resource and scheduling management in cloud computing application paradigm. Paper published at the book "Cloud Computing: Methods and Practical Approaches", Eds. Prof. Zaigham Mahmood, published in Methods and Practical Approaches Series: Computer Communications and Networks by Springer International Publishing, May 13, 2013, ISBN 978-1-4471-5106-7

13. Mavromoustakis C, Bani-Yassein M-M (2012) Movement synchronization for improving file-sharing efficiency using bi-directional recursive data-replication in vehicular P2P systems. Int J Adv Networks Services 5(1 & 2):78–90

14. Chun B-G, Maniatis P (2009) Augmented smartphone applications through clone cloud execution. Paper published at the HotOS

15. Satyanarayanan M, Kozuch M-A, Helfrich C-J, O'Hallaron D-R (2005) Towards seamless mobility on pervasive hardware. Paper published at the Journal of Pervasive and Mobile Computing

16. Satyanarayanan M, Caceres B-P, Davies N (2009) The case for vm-based cloudlets in mobile computing. Pervasive Comput 8(4)

17. Vishwanath V-K, Nagappan, N (2010) Characterizing cloud computing hardware reliability. Paper presented at the 1st ACM Symposium on Cloud Computing, June, pp 193–204

18. Wiboonrat M (2008) An empirical study on data center system failure diagnosis. Paper presented at the 3rd International Conference on Internet Monitoring and Protection, June, pp 103–108

19. Moschakis I-A, Karatza H-D (2012) Parallel job scheduling on a dynamic cloud model with variable workload and active balancing. Paper presented at the 16th Panhellenic Conference, May 2012, pp 93–98

20. Mastorakis G, Mavromoustakis C-X, Bourdena A, Kormentzas G, Pallis E (2013) Maximizing energy conservation in a centralized cognitive radio network architecture. Paper presented at the 18th IEEE International Workshop on Computer Aided Modeling Analysis and Design of Communication Links and Networks (CAMAD), Berlin, Germany, 25–27 September 2013

21. Mastorakis G, Bourdena A, Mavromoustakis C-X, Pallis E, Kormentzas G (2013) An energy-efficient routing protocol for ad-hoc cognitive radio networks. Paper presented at the Future Network & MobileSummit 2013, 3–5 July 2013, Lisbon, Portugal, ISBN: 978-1-905824-36-6

22. Shu S, Jeon W-J, Nahrstedt K, Campbell R-H (2009) Real-time remote rendering of 3D video for mobile devices. Paper presented at the 17th ACM international conference on Multimedia, pp 391–400

23. Dan M., Zhu W, Luo C, Chen C-W (2011) Resource allocation for cloud-based free viewpoint video rendering for mobile phones. Paper presented at the 19th ACM international conference on Multimedia, pp 1237–1240. ACM

24. Weidong S, Lu Y, Li Z, Engelsma J (2011) SHARC: a scalable 3D graphics virtual appliance delivery framework in cloud. J Network Comput Appl 34(4):1078–1087

25. Jouini O (2012) Analysis of a last come first served queueing system with customer abandonment. Comput Oper Res 39:3040–3045

26. Mavromoustakis C-X, Dimitriou C-D, Mastorakis G (2012) Using real-time backward traffic difference estimation for energy conservation in wireless devices. Paper presented at the 4th International Conference on Advances in P2P Systems, Sept. 23–28, 2012, Barcelona, Spain

27. Dimitriou C, Mavromoustakis C-X, Mastorakis G, Pallis E (2013) On the performance response of delay-bounded energy-aware bandwidth allocation scheme in wireless networks. Paper presented at the IEEE ICC2013, 9–13 June 2013, Budapest, Hungary

Chapter 8
Cooperative Strategies for End-to-End Energy Saving and QoS Control

Evariste Logota, Firooz B. Saghezchi, Hugo Marques, and Jonathan Rodriguez

Abstract Energy efficiency and Quality of Service (QoS) have become major requirements in the research and commercial community to develop green communication technologies for cost-effective and seamless convergence of all services (e.g., data, 3D media, Haptics, etc.) over the Internet. In particular, efforts in wireless networks demonstrate that energy saving can be achieved through cooperative communication techniques such as multihop communications or cooperative relaying. Game-theoretic techniques are known to analyze interactions between collaborating entities in which each player can dynamically adopt a strategy that maximizes the number of bits successfully transmitted per unit of energy consumed, contributing to the overall optimization of the network in a distributed fashion. As for the core networks, recent findings claimed that resource over-provisioning is promising since it allows for dynamically booking more resources in advance, and multiple service requests can be admitted in a network without incurring the traditional per-flow signaling, while guaranteeing differentiated QoS. Indeed, heavy control signaling load has recently raised unprecedented concerns due to the related undue processing overhead in terms of energy, CPU and memory consumption. While cooperative communication and resource over-provisioning have been researched for many years, the former focuses on wireless access and the latter on the wired core networks only. Therefore, this chapter investigates existing solutions in these fields and proposes new design approach and guidelines to integrate both access and core technologies in such a way as to provide a scalable and energy-efficient support for end-to-end QoS-enabled communication control. This is of paramount importance to ensure rapid development of attractive 3D media streaming in the current and future Internet.

Hugo Marques is also with the Instituto Politécnico de Castelo Branco, Portugal

E. Logota (✉) • F.B. Saghezchi • J. Rodriguez
Instituto de Telecomunicações, Aveiro, Portugal
e-mail: logota@av.it.pt; firooz@av.it.pt; jonathan@av.it.pt

H. Marques
Instituto Politécnico de Castelo Branco, Castelo Branco, Portugal
e-mail: hugo.marques@av.it.pt

© Springer Science+Business Media New York 2015
A. Kondoz, T. Dagiuklas (eds.), *Novel 3D Media Technologies*,
DOI 10.1007/978-1-4939-2026-6_8

135

8.1 End-to-End Energy Saving and QoS Control Requirements

Energy efficiency is essential for sustainable development of the Information and Communications Technology (ICT), ensuring the profitability of the operators by reducing the Operational Expenditure (OPEX) and prolonging the battery lifetime of mobile User Equipments (UEs). The ICT sector, including wireless and wired networks, is responsible for approximately 2 % of global greenhouse gas emission, which is equivalent to the contribution of the aviation industry [1]. This contribution will increase even rapidly over the coming years with ubiquitous provisioning of ultra-broadband mobile Internet and the explosion of the mobile traffic. In fact, ICT has become an integral part of our lives, playing a key role in modern economy, from e-businesses, teleworking and transportation to health, security and safety. According to Cisco forecast, the global IP traffic has increased fivefold over the past 5 years and will increase threefold over the next 5 years, while video communication will continue to be in the range of 80–90 % of total IP traffic [2]. It is also worth mentioning that mobile traffic almost doubles every year [3]. Hence, if no action is taken, the carbon footprint of the ICT industry is projected to double by 2020 [4]. Therefore, pursuing green communication solutions to reduce the energy cost for transferring one unit of information (i.e., Joule/bit) is of paramount importance to reduce the carbon footprint of not only the ICT sector itself, but also the correlated sectors, such as transportation and safety. As the OPEX is reduced, operators can keep offering affordable services despite the 10–100 times data rate improvement expected by 2020. However, these requirements are very challenging due to the unpredictability of network behaviour as the traditional networking design approaches have shown serious limitations to meet the expectations. Moreover, an effective support for energy efficiency in a network must be end-to-end driven and Quality of Service (QoS)-aware; that is, from the access network to the core network while guaranteeing acceptable QoS delivery to the network users.

From access network standpoint, pursuing energy efficient communication techniques is important for its impact on the battery lifetime of mobile UEs constrained by limited battery. The evolution of the battery capacity for mobile UEs is slow, comparing to the Moore's law, while the traffic generated and consumed by mobile users is growing exponentially. This creates a growing gap between the energy consumption of the UEs and the energy available in the batteries. In fact, unless proper action is taken, there is a threat that mobile users will likely be searching for power outlets than a network access, and once again they will be trapped into a single location—the proximity of power outlets [5]. Hence, energy efficiency has become a key requirement for designing the fifth Generation cellular system (5G), expected to be commercialized by 2020. In fact, for 5G, we need 1,000 times improvement in the energy efficiency of the Radio Access Network (RAN) as well as ten times improvement in the battery lifetime of the UEs [6]. To achieve these energy efficiency requirements, cooperative communications approach is promising. Mobile UEs can adopt cooperative relaying strategy exploiting their spatial diversity to establish radio links that are more reliable and can support higher data rates [7]. For example, the communication can be performed through two independent links, namely direct and relay links. Outage event occurs when both of these links are in deep fade, an event that is unlikely to happen. Multihop communication can also be adopted to overcome channel fading and improve the network coverage [8]. As a UE moves around, the wireless channel quality may deteriorate and at some point the channel quality drops to a level that is unable to support the required QoS. Or, the energy cost of the wireless link (Joule/bit) becomes too expensive, justifying looking for an alternative energy efficient path. As such, alternative links that are able to support the required QoS may be identified, and among them the most energy efficient path may be replaced with the existing one. This can be done by performing (horizontal) handover from one base station to another or vertical handover from one RAN to another in a heterogeneous wireless network.

While cooperative communication promises energy saving for the UEs, which is critical for users' true mobile experience, it imposes dynamic changes of communication paths between the UEs. As we referred earlier, the paths must guarantee sufficient bandwidth to improve users' experience in the dynamic network environment. Indeed, the rapid growth of video demand combined with users' expectations for high QoS delivery has become a constant source of concern for network operators as the network is often blamed for poor quality and video stalling. The video stalling occurs mainly due to the lack of sufficient bandwidth guarantee on the communication paths assigned to the services [9]. To this end, the Internet Engineering Task Force (IETF) developed the Integrated Services (IntServ) [10] and allowed for reserving appropriate amount of bandwidth for each service on each network node (e.g., routers) on the communications paths to assure end-to-end QoS delivery for the services individually. Hence, when a service terminates, the resource reserved for it must be released for future use. The key issue here is that resource reservation, usually resorting to the Resource Reservation Protocol (RSVP) [11], involves signaling to enforce the control policies on the nodes on the path and the related processing consumes CPU (Central Processor Unit), energy and memory [12]. As it is argued in [13], excessive control states, signaling, long call setup time and the related processing overhead are "Achiles' heel" to meet energy efficiency, QoS and scalability targets in the future networks. Therefore, frequent change of communication paths driven by cooperative communication strategies deserves careful attention to prevent undue energy consumption that the QoS enforcement processing may place throughout the network.

To reduce the performance issues raised by the IntServ signaling operations, which are triggered upon every service request or path change, IETF introduced the Differentiated Services (DiffServ) [14], as being a Class of Service (CoS) based QoS architecture standard for the Internet. In DiffServ, the network resource is assigned to each CoS in a static manner (e.g., a percentage of the link capacity), which shows poor resource utilisation as traffic behaviours are dynamic and mostly unpredictable. Therefore, resource reservation must be carried out dynamically, taking into account network's current resource conditions and changing traffic requirements to optimize resource utilization. In this sense, IETF proposed the aggregate resource reservation [15] protocol to allow for reserving more resource than a CoS may need currently, the resource over-reservation. In this way, several requests can be processed without instant signaling as long as previous reservation surplus is sufficient to accommodate new requests [16, 17]. However, the approach imposes a trade-off between signaling overhead reduction and waste of resources including QoS violations [18]. The waste occurs when residual resources (over-reserved but unused) cannot be properly reused, while QoS violations happen when wrong admission decisions lead to accepting more requests than a reservation can accommodate. In view of this, recent findings [19] claimed that resource over-reservation solution can avoid waste by implementing adequate networking architecture and resource distribution protocol that are able to monitor network resource conditions without signaling the network unnecessarily.

Furthermore, it is broadly accepted in the research community that the Internet design needs urgent reconsideration. Many proposals [20] including "clean slate" approach have been put forward. The Open Networking Foundation (ONF), a non-profit and mutually beneficial trade organization, was founded to improve networking performance through the Software Defined Networking (SDN) [21] and its OpenFlow protocol [22] standard. SDN is a networking paradigm which consists of decoupling the control plane (software that controls network behaviour) from the data plane (the devices that forward traffic). The main idea is to make networking control and the management flexible, so one can build the network in many different ways by programming the control logic in terms of architectures, protocols and policies through the control plane. Hence, the SDN deployment cannot simply rely on the traditional control algorithms or solutions to optimize QoS and energy saving performance in a scalable manner.

Therefore, this chapter aims to provide a cooperative communication approach that is able to integrate with advanced resource over-reservation techniques [23] in such a way as to improve energy efficiency, QoS and scalability in an end-to-end fashion. In particular, an SDN controller will be deployed in service provider's network to assure a proper control of the provider's overall infrastructure on one hand and the cooperative communication behaviours on the other hand. This will allow taking into account not only the access network and users' context information, but also the impact of cooperative communication on the control plane in such a way as to optimize the overall performance. For example, a user may not be handed over to another base station unless it ensures better performance in terms of both energy efficiency and QoS guarantee on end-to-end basis, depending on the SDN local control policies. Hence, there is a big potential for further energy efficiency improvement by integrating the wired and the wireless networks and centralizing the resource management of the whole network in a single SDN entity.

This chapter is organized as follows. Section 8.2 explores cooperative communications with focus on the most related work and proof-of-concept for energy saving. Section 8.3 focuses on QoS control mechanisms and provides the most related work and proof-of-concept for efficient signaling overhead reduction. Then, Sect. 8.4 describes our novel approach and key guidelines for efficient integration of cooperative communications and resource over-reservation, aiming at end-to-end energy saving and QoS support towards the future network control designs. Finally, Sect. 8.5 concludes the chapter.

8.2   Cooperative Communications

Statistical variation of the wireless channel due to multipath fading is a major impediment for a reliable wireless communication. Furthermore, the loss of a wireless channel depends on the distance between the transmitter and the receiver, among others. In a dynamic wireless environment, as mobile nodes move around, the distance between the transmitter and the receiver varies, which results in the variation of the underlying wireless channel qualities. As a result, the transmitter has to slow down the data rate frequently in order to cope with the channel variations. The channel fading can even be so deep that no reliable communication is possible, leading to an outage event. As illustrated by Fig. 8.1, User Equipment 1 (UE1) is initially connected to the network through evolved Node B1 (eNB1) and is consuming some 3D media content from a server located in the core network, through the Edge Router 1 (ER1).

As UE1 moves around, the energy cost of the wireless link (Joule/bit) to eNB1 becomes too expensive or the channel quality drops to a level that is unable to support the required QoS. Besides, UE2 which is located in short distance to UE1 may have good channel quality to eNB2. As a result, UE1 can establish a cooperative handover from the existing direct link to eNB1 to a cooperative link to eNB2, through UE2. Having performed the cooperative handover, UE1 continues to download the rest of the 3D media content through the new path (i.e., through ER2).

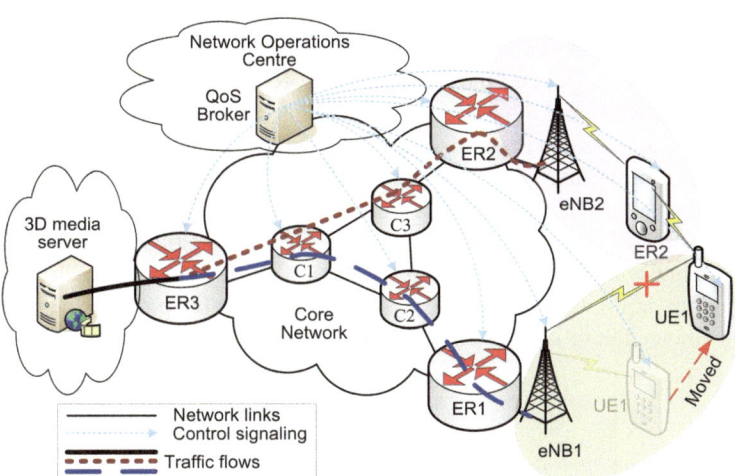

Fig. 8.1 Illustration of users and networks dynamics

8.2.1 Related Work

Cooperative communication has gained considerable interest from both academia and standardization bodies. The existing literature and standards on cooperative communication can be classified into two categories: multihop communication and cooperative relaying. In this section, we elaborate these two categories along with reviewing the related literature. The first form of cooperative communication is multihop communication, primarily employed in ad-hoc networks to ensure connectivity in the absence of a communication infrastructure. In this form of communication, wireless nodes cooperate by forwarding data packets for each other. As such, two nodes that are out of their radio coverage are still able to communicate through some intermediate relay nodes acting as routers (see Fig. 8.2). The IETF Mobile Ad-hoc Networks (MANET) working group develops standards for IP routing protocol suitable for wireless routing application within both static and dynamic topologies [24]. The second form of cooperative communication is cooperative relaying. In contrast to multihop communication, the main motivation for this form of cooperation is not connectivity. Rather, it is employed to exploit the underlying spatial diversity of the wireless channel to enhance the system performance including the link reliability, data rate and energy efficiency.

Figure 8.2 depicts the two above-mentioned types of cooperative communication. The main difference between these two types is: in multihop relaying (the left diagram), there is no direct link between the source node S and the destination node D, so the communication between these two nodes has to be performed through the relay node R. In contrast, in cooperative relaying (the right diagram), there exists a direct link between the source and the destination nodes, and the relay path serves as an alternative path to add more degrees-of-freedom to the channel (spatial diversity). Figure 8.3 shows a snapshot of the signal-to-noise ratio (SNR) variations of the direct path <S–D> and the relay path <S–R–D>. The variations are mainly due to the shadowing and the multipath fading impairments. As seen in the figure, there are time intervals that the received SNR drops below the threshold for a reliable detection; these time intervals have been illustrated by the light-shaded regions in the figure. As the variations of the perceived SNR from the two alternative paths <S–D> and <S–R–D> at the destination node D are statistically independent, the probability that both links are in deep fading is very low, illustrated by the dark-shaded regions in the figure. Therefore, exploiting the relay path besides the direct path can enhance the reliability of the wireless channel considerably.

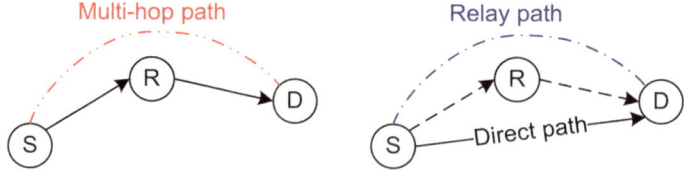

Fig. 8.2 Contrasting multi-hop communication (*left*) and cooperative relaying (*right*)

Fig. 8.3 Time variations of the signal to noise ratio (SNR) for the direct and the relay paths

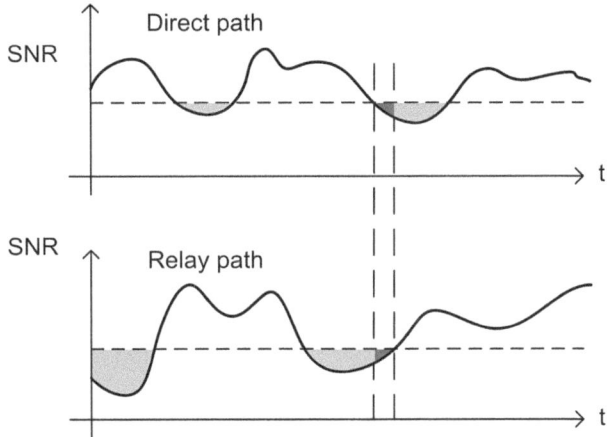

Cooperative relaying has been investigated to improve the spectral-efficiency (SE) of the wireless system. The application of cooperative relaying in wireless systems can be traced back to the groundbreaking work of Cover and El Gamal [25] on the information-theoretic characterization of the relay channel. The work builds upon the three-node (one source node, one destination node and one relay node) channel model first introduced by Van der Meulen [26] and examines its channel capacity for the case where the channel is contaminated by additive white Gaussian noise (AWGN). Recent works on this subject are mainly focused on taking advantage of the underlying spatial diversity introduced by the relay node(s) to combat channel fading and improve the reliability of the system [27–37]. The spatial diversity has also been exploited to improve the spectral efficiency of the wireless channel through distributed space-time multiplexing techniques [38, 39]. One possibility of cooperative relaying is that mobile terminals can share their antennas and form a virtual multiple-input–multiple-output (MIMO) channel to take advantage of the provided spatial diversity. In contrast to the transmit diversity or the receive diversity, this form of spatial diversity is referred to as *cooperative diversity*. Laneman et al. [27] developed different cooperative fixed relaying schemes such as Amplify-and-Forward (AF) and Decode-and-Forward (DF), selection relaying and incremental relaying, and examined their performance in terms of the outage probability. Unlike this work where the authors constrained the relay node to operate in the half duplex mode, employing TDMA scheme, Sendonaris et al. [29, 30] studied the cooperation of mobile users when both users have data to transmit and address practical implementation issues within a CDMA framework.

Cooperative communication can also be adopted in cellular systems to improve the system efficiency. The resulting network from the integration of multihop communication with cellular system is referred to as a Multihop Cellular Network (MCN) or a Hybrid Ad-hoc Network (HANET) [40–42]. There has been considerable interest from both the standardization bodies and the academia in MCNs. Opportunity Driven Multiple Access (ODMA) [43] is a multihop relaying protocol, proposed by the 3rd Generation Partnership Project (3GPP) to be applied to

Universal Mobile Telecommunications System (UMTS) to improve the network capacity, the SE at cell edges and the coverage at hotspots. Relay technology has also been standardized by 3GPP in Release 9 and Release 10 of the Long Term Evolution-Advanced (LTE-A) [44–46], where relays are deployed either at fixed locations or on mobile vehicles (e.g., trains, buses, etc.) to extend the coverage and enhance the spectral- and the energy-efficiency. A relay node is attached to an eNB through a wireless backhaul link either using the same frequency as the one used in the access part (inband relaying or self-backhauling) or using a different frequency band (outband relaying). 3GPP also defines different types of relays functioning in different network layers varying from layer 1 (PHY layer) to layer 3 (Network layer) [47]. Layer 1 relay is the simplest form of the LTE relay functioning in AF mode. Layer 2 and Layer 3 relays operate in DF mode. A Layer 3 relay is indeed a small eNB with the full functionality such as transmitting control signaling or performing mobility management. As for the academic activities on the MCN, in [48], the authors proposed the integration of the Mobile Ad-hoc Network (MANET) and the Global System for Mobile (GSM) and introduced a so called Ad-hoc GSM (A-GSM) platform, addressing practical issues for evolutionary changes of the GSM system in order to support the relaying functionality for the voice calls. Additionally, in [49], the authors overview several contributions to Working Group 4 of the Wireless World Research Forum (WWRF) and present several relay-based deployment concepts such as multihop relaying, cooperative relaying, virtual antenna arrays and so on.

Recently, cooperative communication has been exploited to extend the battery lifetime of UEs. C2POWER [50] and Green-T [51] are two European research projects aiming at exploiting cooperative and cognitive communication to reduce the power consumption of the multi-mode UEs. In [52, 53], the authors study the energy saving performance of cooperative communication exploiting short range (SR) interfaces of the multi-mode UEs. Interacting as independent agents while cooperating or competing for efficient resource utilization, conflict of interest naturally occurs among the mobile users. Game theory is a mathematical tool that can be applied to analyse strategic interaction among rational players [54–63]. Specifically, with the adoption of cooperative communication in wireless networks, coalitional game theory (a.k.a. cooperative game theory) can be applied to incentivize UEs to cooperate with each other [56–59]. Solution of a non-cooperative game—which is given by the well-known Nash Equilibrium—basically predicts the equilibrium strategy profile where no player wants to change its strategy. On the other hand, solving a cooperative game means incentivizing all players by dividing the common payoff that the players obtain by their cooperation in a fair way where all players are satisfied and no one has incentive to leave the cooperation. Further detail about cooperative game theory is out of the scope of this chapter. The interested is referred to [59].

Another important issue related to the cooperation of UEs is detecting and isolating selfish nodes from the cooperative ones. There are two main techniques for this purpose, namely credit-based schemes [64–67] and reputation-based schemes [68–73]. In the former, there is a central trustworthy entity that maintains a credit account for each UE, where the cooperation of the UEs is fully recorded and can be tracked. A UE gains credit when cooperates and spends credit to receive help from other UEs; the spent credit is transferred to the account of the helping UE (i.e. the UE that is acting as a relay). Therefore, cooperative UEs increase their credit level over time while the selfish ones fail to accumulate any credit in their account. If a selfish node, which has no credit in its account, requests for a packet forwarding service from the cooperative UEs, they will refuse the request. As a result, selfish nodes are gradually excluded from the cooperative groups. In contrast to the credit-based schemes, which are based on a centralized approach, the reputation-based schemes are based on a distributed trust management approach, where every mobile user maintains a reputation record of its neighbouring UEs. The UEs also disseminate the reputation image of their immediate neighbour UEs to the other UEs located far away to accelerate the detection and isolation of the selfish UEs. It is worth pointing out that there is an intricate difference between the credit-based schemes and the cooperative game-theoretic techniques: in the former, normally players receive flat credit regardless of their effort for or their impact on the achieved energy saving, while, in the later, the players receive credit proportional to their influence and effort.

8.2.2   Proof-of-Concept for Energy Saving

For proof-of-concept, we perform a simulation to evaluate the impact of cooperative communication on the battery lifetime of the UEs. Figure 8.4 illustrates the considered scenario, where several UEs under the radio coverage of a WiFi Access Point (AP) make a coalition and cooperate to save energy. We assume that every UE is equipped with two radio interfaces, namely WiFi (IEEE 802.11g) and WiMedia.

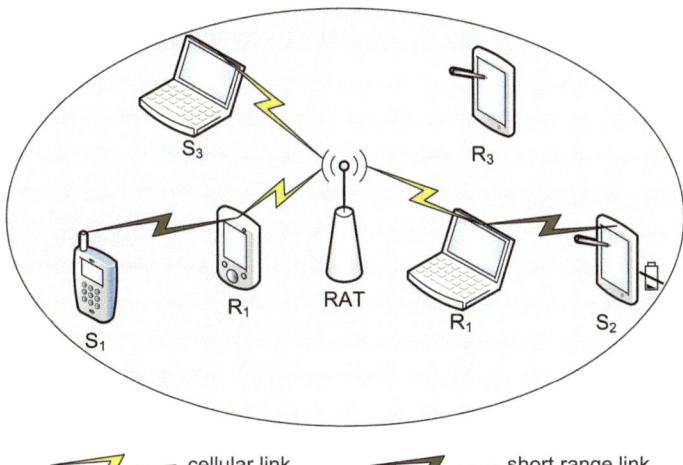

Fig. 8.4 A coalition of UEs cooperating to save energy [75]

The WiFi interface is used to communicate with the AP, while the WiMedia interface is used for short range communications. As illustrated by the figure, there are three source nodes (S_1, S_2 and S_3) and three relay nodes (R_1, R_2 and R_3). Source nodes communicate with the AP over either a conventional direct link or a cooperative short range link. A cooperative short range link is a two-hop link—with only one intermediate relay node between the source and the destination. Communication over the first hop (from the source to the relay) is performed over a short range WiMedia link, while the communication over the second hop (from the relay to the AP) is performed over a WiFi link.

Exploiting their multiple radio interfaces, the relays operate in a full-duplex mode. They receive a packet through their short range WiMedia interface and decode it. Then, they encode and forward it through their long range WiFi interface to the AP. In the scenario shown by Fig. 8.4, source node S_1, which is experiencing a bad channel quality to the AP, cooperates with relay node R_1 that has good channel quality to the AP; source node S_2, which is running on low battery, finds a nearby relay R_1 with good battery level to cooperate with; and finally source node S_3, which has a good channel quality to the AP, communicates with the AP directly. Relay node R_3, despite willing to cooperate, is left unused as its cooperation brings no additional energy saving to the coalition. We conduct the simulation for different coalition sizes (ranging from 10 to 100 nodes) and different relay densities (low density of 20 %, medium density of 50 % and high density of 80 %). For example, if the coalition size is ten, a relay density of 20 % means that only two of the UEs in the coalition are relay nodes, while the remaining eight UEs are acting as source nodes. During the simulation, which lasts for 100 s, every source node generates a constant bit rate (CBR) traffic with the rate of 10 packets per second and the packet size of 1024 bytes. We use channel models and system parameters summarized in [74] as well as random way point (RWP) mobility model with the maximum speed of 3 m/s and the pause time of 5 s for the mobile nodes.

Figure 8.5 illustrates the average energy saving gain (ESG) of the coalition when the coalition size varies between 10 and 100 UEs for three different densities of the

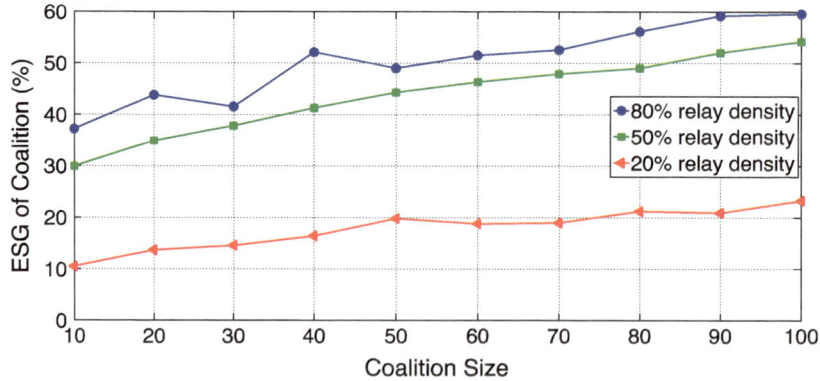

Fig. 8.5 Energy saving gain (ESG) for different coalition sizes and relay densities [59]

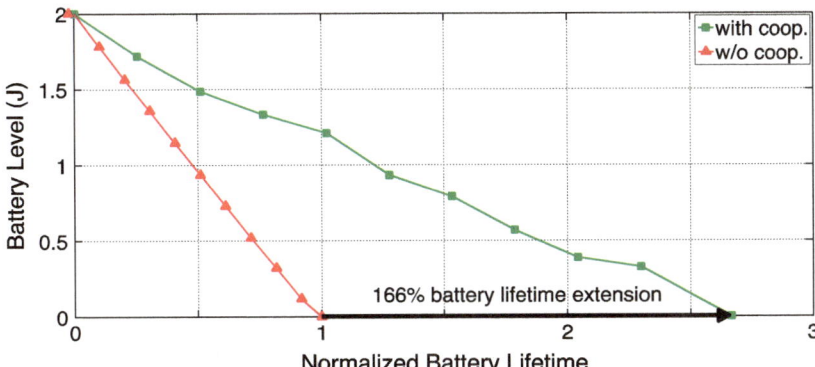

Fig. 8.6 Battery lifetime extension of the UEs adopting cooperative communication [74]

relays, namely 20, 50 and 80 %. As seen from the figure, the ESG increases when the coalition size or the relay density increases. This is due to the fact that the cooperation opportunity increases among UEs as the coalition size or the relay density increases. We also observe that when the relay density increases from 20 to 50 %, the ESG increases significantly, but it increases marginally when the density of relays saturates, while further increasing from 50 to 80 %. Finally, as seen from the figure even with a moderate coalition size of 10 UEs and the relay density of 50 %, it is possible to reduce the energy consumption of the coalition by 30 %.

To study the impact of cooperative communication on the battery lifetime of the UEs, we conduct a simulation with 50 UEs, 25 UEs acting as source nodes and the rest 25 UEs acting as relay nodes. Every UE starts with an initial battery level of 2 Joules and an initial credit level of 0.1. The simulation lasts for 600 s, during which the role of the sources and the relays are switched around every 5 s. To benchmark, we repeat the simulation for the case when there is no cooperation and all the short range interfaces are switched off and all communications are performed over conventional direct links. Figure 8.6 shows the battery level of the UE that has the minimum battery level among the UEs. The figure contrasts the battery lifetime extension of this UE against the benchmark case, where there is no cooperation. As seen from the figure, cooperation can extend the battery lifetime of the UEs up to 166 %—more than 2.5 times. Further discussion on the battery lifetime extension of the UEs using short range cooperation technique is available in [59, 74].

To incentivize the UEs to cooperate, we apply a credit scheme based on cooperative game theory. We assume that there is a virtual central bank (VCB) that maintains an account for every UE where the history of its cooperation can be traced and recorded. We model the problem as a cooperative game and award cooperative nodes energy credit. Specifically, we distribute the achieved energy saving in the coalition, using the *core solution* of the cooperative game [75].

A nice property of this approach is that the cooperative UEs receive credit based on their contribution to the achieved energy saving, satisfying all the UEs and incentivizing all of them to keep the cooperation on. Last but not least, when a UE acts as a relay, it consumes its energy to forward packets for the other UEs. Unless this energy consumption is reimbursed, the relay will have no motivation to keep on acting cooperatively. To this end, we provide the relay some additional credit, the amount of which is equal to its consumed energy for the packet forwarding. Obviously, this is apart from the credit that it receives from the distribution of the achieved energy saving (determined by the *core solution*) explained previously. Similarly, we charge the source nodes for receiving packet-forwarding service from the other UEs. Therefore, our proposed credit scheme serves not only to ensure a sustainable cooperation of the UEs, but also to detect and isolate the selfish UEs from the cooperative ones.

In order to evaluate the effectiveness of the proposed credit scheme to detect and isolate the selfish players, we conduct another simulation with a coalition size of 50 UEs and 50 % relay density (25 sources and 25 relays). Among them, 5 UEs are selfish, while the rest 45 UEs are cooperative. The simulation starts with all the UEs having equal initial battery of 2 Joules and an initial credit level of 0.1. Similar to the previous simulation, every source UE generates a CBR traffic with the rate 10 packets per second and the packet size of 1024 bytes. Similar to the previous simulation, every 10s, the sources and the relays change their roles, giving chance to every UE to act both as a relay and as a source equally likely. The simulation lasts until all UEs run out of battery. Figure 8.7 compares the average credit level of the cooperative and the selfish UEs. As seen from the figure, all the UEs start with an equal initial credit level of 0.1. The cooperative UEs gain credit and increase their credit level, while the selfish ones lose their initial credit soon. As soon as a UE is left without credit, other UEs avoid cooperating with it, which leads to its isolation from other UEs. Further discussion on this topic is out of the scope of this chapter. The interested reader is referred to [75].

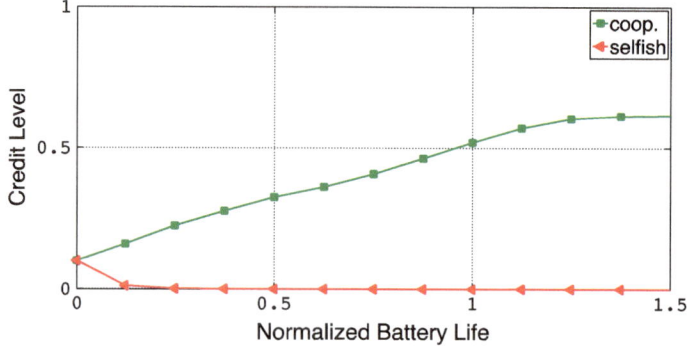

Fig. 8.7 Comparing credit level of the cooperative and the selfish UEs [75]

8.3 QoS Provisioning in Packet Switched Network

The IP network convergence [76] has imposed the Internet as being the core infrastructure to attach various access technologies and to assure the transport of all kind of services and applications (e.g., data, audio and video). A major challenge in the convergence environment is that each service or application has its own QoS requirements for bandwidth, latency, jitter and loss. For example, applications like Voice over IP (VoIP) require 150 ms of (mouth-to-ear) delay, 30 ms of jitter and no more than 1 % packet loss [77]. The interactive video or video conferencing streams embed voice call, and thus have the same service level requirements as VoIP. In contrast, the streaming video services, also known as video on-demand, have less stringent requirements than the VoIP due to buffering techniques usually built in the applications. Other services such as File Transfer Protocol (FTP) and e-mail are relatively non-interactive and delay-insensitive. This means that the Internet must be able to treat each service according to the service requirements and efficiently utilize the network resources for operators to maximize revenue. However, the legacy Internet system was not designed for these purposes; it treats all the services equally in a best-effort fashion. In order to make the Internet more attractive, the QoS provisioning consists of designing algorithms and protocols to enable the network to provide predictable, measurable and differentiated levels of quality guarantees. This is usually enabled through service admission control, service prioritization and network resource reservation. The resource reservation, usually resorting to the Resource Reservation Protocol (RSVP) [11], allows reserving appropriate amount of bandwidth on the path to be taken by the packets that belong to the service requiring the bandwidth. Further details are provided in the subsequent subsection.

8.3.1 Related Work

Network resource reservation control has become one of the most exciting frontiers of research and development in the communications systems to deliver guaranteed QoS. Key relevant standards of the QoS approaches include the IP Multimedia Subsystem (IMS) from the 3GPP, the Resource Admission Control Function (RACF) from the ITU-T (International Telecommunication Union Standardization Sector) and the Resource and Admission Control Sub-system (RACS) from the TISPAN (Telecommunications and Internet converged Services and Protocols for Advanced Networking) [13]. Traditionally, the resource reservation is performed on per-flow basis, meaning that the QoS signaling messages are triggered upon every service request [10]. As a consequence, the approach has been criticized in the research community due to the lack of scalability and therefore the energy saving issues [12]. To ease the understanding, Fig. 8.8 illustrates the processes related to QoS-enabled path setup for establishing a new service session, the path

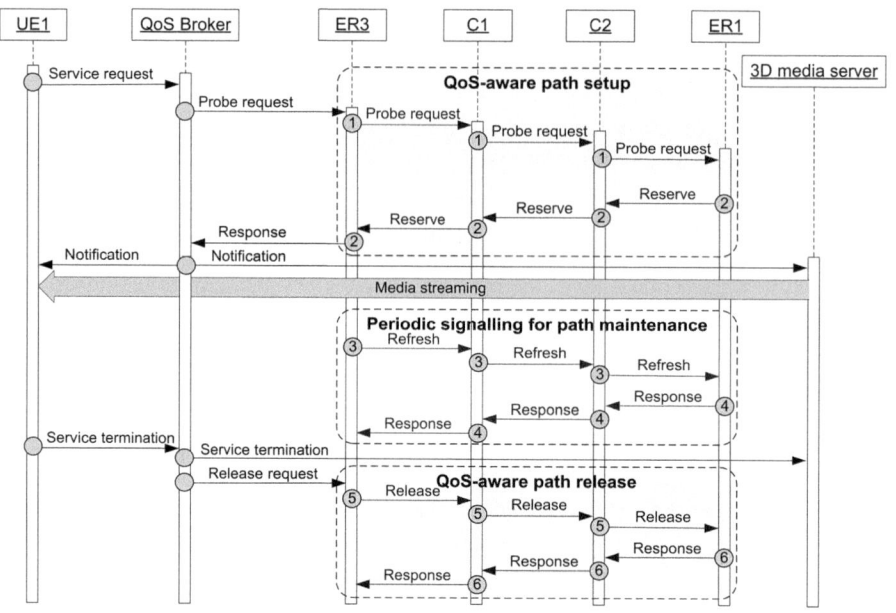

Fig. 8.8 Per-flow QoS-aware communication path control

maintenance during the session lifetime and the path release when the session terminates. Hence, the UE1 (see Fig. 8.1) wants to use the 3D media service from the remote media server. So, UE1 sends a service request to the QoS Broker located in the Network Operations Centre. Note that the QoS Broker is responsible for the overall control of the network. In this sense, the QoS Broker triggers the Edge Router 3 (ER3) to check whether the available network resource is sufficient to transport the requested 3D media with acceptable quality. For this purpose, ER3 sends a Probe message on a path, which may be provided by the underlying routing protocol, Open Shortest Path First (OSPF) [86], to acquire the minimum unused resource on the path. Assuming that there is sufficient available resource, the receiving ER1 signals the path in the reverse direction, so the required resource is reserved on each node on the path and the session is established afterward. Otherwise, the request must be denied or admitted in best-effort CoS, depending on the Service Level Agreement (SLA) between the provider and the user. Then, as long as the 3D media session is active (running), the Refresh and Response signaling messages are periodically (e.g., every 30 s) issued in such a way as to maintain the reservation parameters on the path. When the 3D media session terminates, the Release and Response messages are used to release the reserved resources for future use.

It becomes clear that the per-flow QoS reservation control is not acceptable from the scalability and the energy efficiency perspectives. This is even more critical in cooperative communications scenarios where the UEs, seeking to maximize their battery lifetime, may happen to dynamically change their access Edge Routers. For example, the UE1 (see Fig. 8.1), previously attached to ER1, may change to the ER2 for better performance at lower energy cost. As a consequence, the path <ER3, C1, C2, ER1> must be released and a new one <ER3, C1, C3, ER2> must be established to properly connect ER3 and ER2, depending on the route provided by the routing protocol in use. Hence, while the cooperative approach promises energy saving in the wireless environment, lack of scalable QoS control solutions can easily increase the energy consumption in the core network. In order to address these challenges, resource over-reservation, consisting of reserving more resources than a CoS may need, must be explored, so several service requests can be processed without instant signaling as long as the reservation surplus is sufficient to accommodate new requests [12, 15, 17]. However, as we explained earlier in Sect. 8.1, the approach imposes a trade-off between signaling overhead reduction and waste of resources as well as QoS violations [18]. ITU-T G.1081 [83] suggests five monitoring points in networks, allowing service providers for monitoring networks and services performance to improve resource utilization. However, existing proposals are mostly base on path probing techniques [84] and show limitations in terms of complexity, accuracy and undue signaling [85].

Pan et al. [87] proposed to over-reserve bandwidth surplus as a multiple of a fixed integer quantity for aggregate flows destined to a certain domain—a Sink-Tree-Based Aggregation Protocol. This solution also does not comply well with network dynamics and fails to efficiently utilize the network resources. The work in [88] over-provisions virtual trunks of aggregate flows based on a predictive algorithm (e.g., past history) without an appropriate mechanism to dynamically control the residual bandwidth between various trunks. Resource over-reservation is also studied in [89] in an attempt to reduce control states and the signaling overhead. Sofia et al. [90] proposed the use of resource over-reservation to reduce excessive QoS signaling load of the Shared-segment Inter-domain Control Aggregation Protocol (SICAP). However, these proposals lead to undesired waste of bandwidth. The Multi-user Aggregated Resource Allocation (MARA) [17] proposed functions to dynamically control bandwidth over-reservation for CoSs to improve system scalability. However, MARA also shows serious limitations in its resource distribution capability.

In this scope, recent findings proposed new ways for scalable, reliable, cost- and energy-efficient control design of IP-based network architectures and protocols [23]. In particular, the Self-Organizing Multiple Edge Nodes (SOMEN) [79] enables multiple distributed network control decision points to exploit network paths correlation patterns and traffic information in the paths (obtained at the network ingresses and egresses) in such a way as to learn network topology and the related links resource statistics in real-time without signaling the paths. As such, SOMEN provides a generic network monitoring mechanism. The Advanced Class-based resource Over-Reservation (ACOR) [19] effectively demonstrated the breakthrough that it is possible to design IP-based networking solutions with significantly reduced control signaling overhead without wasting resources or violating the contracted quality in the Internet. The Extended-ACOR (E-ACOR) [81] advances the ACOR's solution through proposals of multi-layer aggregation of resource management and a new protocol to efficiently track congestion information on the bottleneck links inside a network without undue signaling load. Nonetheless, to the best of our knowledge, there exists no solution that leverages these advanced over-reservation techniques jointly with cooperative communication strategies in RAN to improve the end-to-end energy efficiency in the network, so further research was deemed necessary.

8.3.2 Proof-of-Concept for QoS Control Signaling Overhead Reduction

As a proof-of-concept for the superiority of resource over-reservation over per-flow approaches in terms of significant reduction of the signaling overhead to achieve scalability, a simulation is performed using the network topology presented in Fig. 8.1 and the Network Simulator (ns-2) [91]. For the sake of simplicity, each network interface is configured with a capacity of 1 Gbps and 4 CoSs: one control CoS (for control packets), one Expedited Forwarding (EF), one Assured Forwarding (AF) and one Best-Effort (BE) [92], under the Weighted Fair Queuing (WFQ) scheduling discipline [93]. In order to simulate a cooperative communication environment with dynamic changes of access points and therefore the paths, 20,000 session requests belonging to three different traffic types, such as CBR, Pareto and Exponential are randomly generated and mapped to various CoSs and ER1 and ER2 based on Poisson processes. The traffic bandwidth requests are generated using uniform distribution between 128 Kbps and 8 Mbps. To show more stable results, we run the simulation five times with different seeds of random mapping of requests to CoSs. Then, the mean values are plotted for all seeds with a confidence interval of 95 %. Further details on the simulation are available in [94].

Figure 8.9 shows that the resource probing and the number of reservation events overlap when the network is less congested (request number below 4,000). As the request number increases further, the number of probing events goes higher than that of the reservation. This means that requests are denied when the probing reveals insufficient resource to guarantee the QoS demanded. The release messages events were also tracked upon session termination and the overall per-flow signalling events number (probing + reservation + release) is plotted in Fig. 8.9. Besides the per-flow results, the resource over-reservation performance is also shown. As one can observe, the over-reservation control effectively allows drastic reduction of the QoS control signaling events. More importantly, no QoS signaling is triggered when the network is less congested since the resource is reserved in advance. The signaling is invoked only when the over-reserved resource parameters need re-adjustment to prevent CoS starvation. Generally, in Fig. 8.9, the over-reservation allows a reduction in the number of signaling events beyond 90 % of that of the per-flow approach, depending on the network congestion level throughout the network. Therefore, resource over-reservation techniques are necessary to ensure

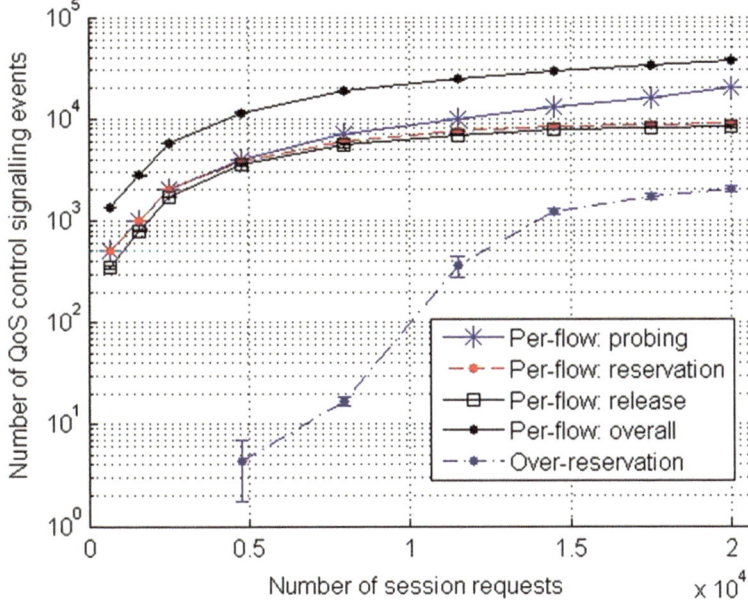

Fig. 8.9 QoS signaling events of per-flow vs. over-reservation control

a viable signaling load (and the related energy consumption), which is of great importance, especially when cooperative communications increase the network dynamics.

While recent QoS over-reservation mechanisms are promising for smart network control designs, they have limited capabilities for being integrated with a heterogeneous and dynamic RAN expected in 5G cellular systems. In addition, their support for SDN and cooperative communication approaches is yet an open research issue that deserves careful attention. Further details and guidelines are provided in the subsequent section.

8.4   Integrating Cooperative Strategies and Advanced QoS Control

This section aims at describing a generic mechanism that is able to exploit SDN Controller (SDNC) to integrate cooperative communications strategies and network resource over-reservation management to save energy while guaranteeing the required QoS. In particular, the SDNC is responsible for granting or denying access to the network and the related resources in order to ensure that: (1) each admitted user receives the contracted QoS; (2) the cooperative communication decisions are managed to enable optimal energy saving across the network. Other functions of the

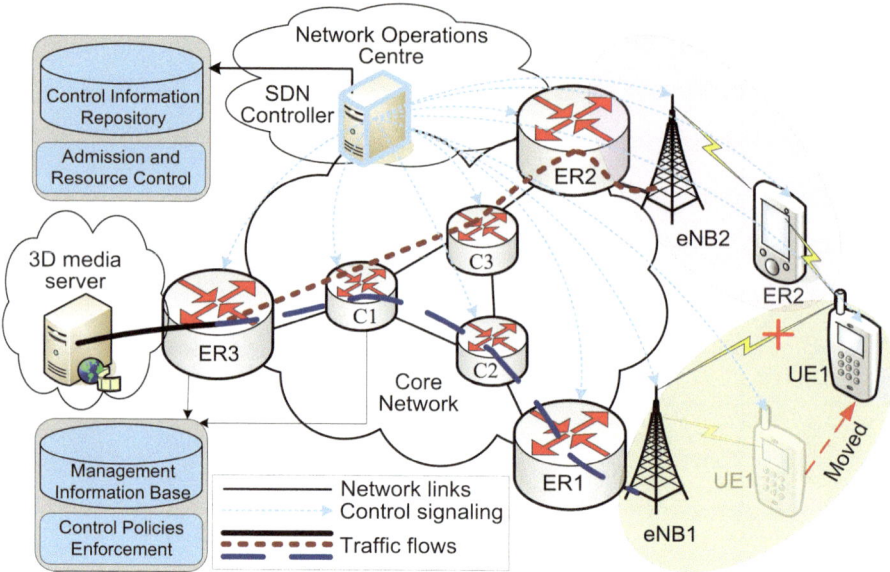

Fig. 8.10 Architecture for energy saving and scalable QoS control scenarios

SDNC include, but not limited to, traffic load balancing to avoid unnecessary congestion occurrence inside the network. In order to achieve this, the SDNC is deployed to define appropriate control policies and dictate the enforcement on the network transport elements (e.g. switches and routers) through appropriate signalling protocols (e.g., OpenFlow protocol [22]). Hence, as illustrated by Fig. 8.10, the SDNC implements two main components: (1) the *Control Information Repository (CIR)* as being a database for maintaining network topological and resource information including the users' profiles; (2) the *Admission and Resource Control (ARC)*, as the entity responsible for defining appropriate control policies and managing the access to the network resources.

Besides, the transport elements implement a *Management Information Base* (MIB) and *Control Policies Enforcement* (CPE) components, as detailed further in subsequent subsection.

8.4.1 Network Control Components and Functionalities

To ease the understanding of the network configurations and operations described later in the Subsec. 8.4.2, this subsection describes the functionalities of the main network control components depicted in Fig. 8.10.

8.4.1.1 Control Information Repository

The CIR is used to maintain the network topology and the resource statistics of the related links, including the communication paths in a real-time manner. The network topology and paths may be pre-defined, dynamically discovered or computed as in [19]. Moreover, the links' resource statistics include the overall capacity of each network interface, and the amount of bandwidth reserved and used in each CoS on the interface [19]. Further, the CIR records the user's sessions-to-CoS mapping along with the related configuration parameters. These parameters include the QoS requirements of the session (e.g., bandwidth, delay, jitter and packets loss), the session ID, the IDs of the flows that compose the session, the ID of the CoS to which the session belongs, the flows' source and destination IDs (e.g., IP and Media Access Control (MAC) addresses) and the ports IDs. Additional users' relevant profile information such as billing and personalization parameters may also be stored in the CIR.

8.4.1.2 Admission and Resource Control

With the ARC component, the SDNC is able to control the access to the network, which spans the coordination of the cooperative UEs within the network. The SDNC achieves this by dynamically taking into account not only the incoming service request QoS requirements (e.g., bandwidth) and the network resource availability obtainable from the CIR database, but also the potential energy efficiency of the adopted strategy. It is responsible for creating and managing the communication paths inside the network. Hence, the ARC provides an interface to receive requests and also to send control instructions to be enforced throughout the network. Whenever there is a session request admission, termination or readjustment of the QoS requirements of a running session, the relevant information (e.g., resource usage) is updated in the CIR in real-time fashion. This is important to maintain a good knowledge of the network resource conditions without undue probing signaling. Another advantage, for example, is that the ARC can exploit the updated information from the CIR to optimize control performance in terms of traffic load-balancing. Moreover, the ARC implements the advanced resource over-reservation algorithm introduced in [19, 23], so session requests can be admitted, terminated or readjusted without instant signaling or resource wastage in the network. Hence, upon exhaustion of the over-reserved resource in a requested CoS on a path, the ARC can re-compute new reservation parameters and trigger the reservation readjustment among the CoSs on the path to prevent the wastage [19, 23].

8.4.1.3 Management Information Base and Control Policies Enforcement

The MIB and the CPE components enable network transport nodes (see Fig. 8.10) to carry out the basic control functions required in all network nodes. In particular, the MIB is a database that can be used to maintain OpenFlow table [22] and key control information to assist packet forwarding and the nodes' reaction to failures. Besides, the CPE component assures interaction with Resource Management Functions (RMF) [96] to properly configure schedulers on the nodes [80] so that each CoS receives the amount of bandwidth assigned to it. The nodes execute these functions based on the instructions that they receive from the ARC. To this end, the CPE enables UDP port recognition (routers are permanently listening on a specific UDP port) or IP Router Alert Option (RAO) [95] on nodes to properly intercept, interpret and process control messages. The CPE also assists the SDNC by filling in relevant control messages with appropriate information such as the IDs (e.g., IP or MAC addresses) of outgoing interfaces and their capacities as *Record Route Object* (RRO) [97, 98] on paths. When the MIB and the CPE are deployed at the network ERs, they include additional functions for inter-domain forwarding and routing operations [99]. Traffic control and traffic conditioning (e.g., traffic shaping and policing) must also be enforced at the ERs to force admitted traffic flows to comply with the SLAs [14].

8.4.2   Network Configurations and Operations

The SDNC can rely on the underlying link state routing protocols [86] to discover and maintain the network topology dynamically. Hence, the network topology is fed to appropriate algorithms (e.g., Dijkstra [86]) to compute all possible paths, especially the ER-to-ER paths inside the core network under the SDNC's control. One may want to combine the ER-to-ER paths to create all possible branched paths for multicasting purposes upon need [17]. Among the possible paths, the best ones that can be used for efficient service delivery can be filtered and maintained based on appropriate criteria such as the number of hops, bottleneck link bandwidth and/or delay. It is worth recalling that the use of deterministic paths, such as the Label Switched Paths (LSPs) in Multiprotocol Label Switching (MPLS), is very important to improve network resource control and the traffic engineering under resource over-reservation. Besides the LSPs, deterministic paths can be maintained also by properly configuring a unique multicast channel (Source, Group) on each path, as in [19]; the Source may be the IP address of the ingress ER where the path originates from and the Group may be a standard multicast address. Further, the ARC computes initial over-reservations parameters for the CoSs at each interface on each path and sends the information to the nodes on the path using the OpenFlow protocol.

As a result, every visited node on the path intercepts the message and configures its local interfaces and the MIB accordingly (e.g., OpenFlow tables, forwarding/ routing tables and resource over-reservation parameters). In this way, the network is initialized and set to run. As the network is operating, every service or cooperative communication request must be directed to the SDNC for the processing. This is to assure service Authentication, Authorization, and Accounting (AAA), QoS and energy efficiency simultaneously to allow optimal network resource utilization. In general, a service request should specify the traffic characteristics (e.g., source and destination IP and port addresses, supported codecs, etc.), the desired QoS (e.g., bandwidth, delay, jitter, packet loss and buffer), the energy requirement and service personalization preferences. This session negotiation process may be handled by using the OpenFlow protocol [22], the Session Initiation Protocol (SIP) [100] or any other signaling protocol (e.g., Next Step In Signaling (NSIS) compliant protocol [97]), specified by the network operator.

For the purpose of exposition, let us suppose that UE1 (previously attached to eNB1) needs to perform handover to eNB2 through a cooperative link with UE2, as in Fig. 8.10. Hence, UE1 issues a service request to the SDNC. The service request contains the context information about UE1 such as its QoS requirements (e.g., data rate, delay, etc.), the traffic characteristics (e.g., codec), the remaining battery level, the list of UE1's available radio interfaces as well as the list of its neighbour UEs (e.g., UE2). Based on the information received, the ARC evaluates the benefits to the end-user and the impact on the overall network performance. Then, it decides to grant or deny the handover request through UE2. More importantly in this approach, the SDNC does not need to signal the communication paths to probe resource availability since the CIR is populated with the information on real-time manner. Also, in case the switching is granted, no QoS reservation release message is needed along the previous path <ER3, C1, C2, ER1>. In addition, assuming that the resource over-reservation solution described in [23] is deployed, as depicted by Fig. 8.11, no QoS resource reservation signaling message is required on the new path <ER3, C1, C3, ER2> as long as the over-reserved resource is sufficient in the requested CoS on the path. As such, it becomes clear that resource over-reservation effectively allows for dynamic switching between QoS-enabled paths without undue signaling overhead, thus contributing to the energy efficiency throughout the network.

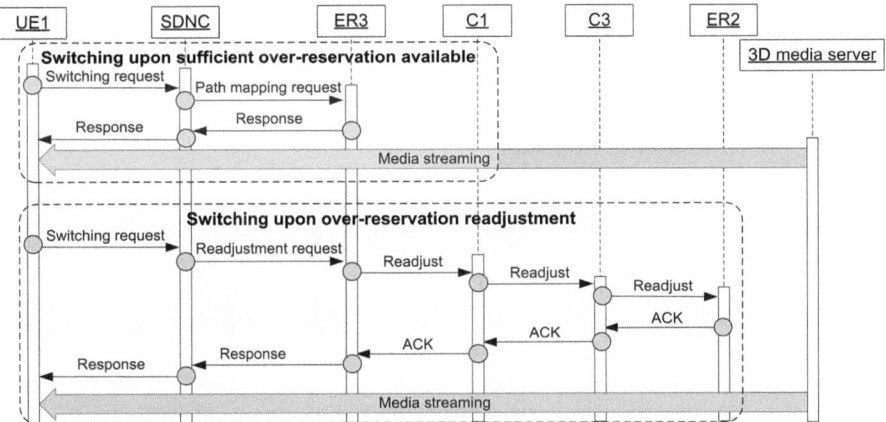

Fig. 8.11 An illustration of QoS over-reservation supporting cooperative communication

In case the over-reserved resource in the requested CoS is insufficient on the path <ER3, C1, C3, ER2>, the ARC detects that QoS reservation readjustment is necessary on the path. Since extra signaling for QoS readjustment on every node on the path would incur processing and communication overhead—which directly translates into additional energy cost, the SDNC must implement appropriate algorithm to ensure overall optimization of the network service performance. As discussed earlier, the SDNC can take into account the energy saving achieved in the RAN and the energy cost spent for the extra signaling in the core network. The SDNC may grànt permission to switch to the new path only when the imposed signaling cost does not outweigh the achieved energy saving.

Finally, the UEs' behaviours are subject to the SDNC's decisions under appropriate control policies. For example, should it be worth to carry out the QoS readjustment and accept the switching (e.g., based on SLA between the operator and the users), the SDNC triggers the operations and the 3D media streaming is switched to the new path in a seamless way to the end-user.

8.5   Summary

This chapter addresses energy efficiency for emerging 3D media applications without compromising the required QoS. We adopt a novel approach and provide guidelines bringing together cooperative communication techniques from RAN and resource over-provisioning techniques from the core network. The proposed approach ensures efficient integration of the RAN and the wired core network in such a way as to enable differentiated QoS support and energy saving in a scalable and end-to-end fashion. In this way, each network end-user will be able to enjoy a minimum guarantee of QoS in terms of bandwidth, delay, jitter and packet loss while experiencing longer battery lifetime achieved through intelligent selection of the partner relay node(s). Moreover, the control allows optimal network resource utilization and drastic reduction of the QoS control signaling overhead, which undoubtedly translates into additional energy saving throughout the network. We believe this approach is essential for the future green mobile and internetworking designs to effectively enhance system overall performance.

Acknowledgments The authors would like to thank the support given by the ROMEO project (grant number: 287896), which was funded by the EC FP7 ICT collaborative research programme.

References

1. Gartner (2007) Gartner Estimates ICT Industry Accounts for 2 Percent of Global CO2 Emissions. http://www.gartner.com/newsroom/id/503867. Accessed 11 July 2014
2. The Zettabyte Era—Trends and Analysis (2014). http://www.cisco.com/en/US/solutions/collateral/ns341/ns525/ns537/ns705/ns827/VNI_Hyperconnectivity_WP.html.   Accessed 30 June 2014

3. Cisco (2010) Cisco visual networking index: global mobile data traffic forecast update, 2010–2015. http://newsroom.cisco.com/dlls/ekits/Cisco_VNI_Global_Mobile_Data_Traf fic_Forecast_2010_2015.pdf. Accessed 9 July 2014
4. Fehske A et al (2011) The global footprint of mobile communications: the ecological and economic perspective. IEEE Commun Mag 49(8):55–62
5. Fitzek F, Katz MD (eds) (2006) Cooperation in wireless networks: principles and applications: real egoistic behavior is to cooperate. Springer, New York
6. GreenTouch (2014) Energy efficient wireless networks beyond 2020. ITU-R Workshop on Research Views on IMT Beyond 2020. http://www.greentouch.org/uploads/GreenTouch_ITU-R_Workshop_Research_Views_on_IMT_Beyond_2020_02-14-2014.pdf.   Accessed: 9 July 2014
7. Dejonghe A et al (2007) Green reconfigurable radio systems. IEEE Signal Process Mag 24 (3):90–101
8. Saghezchi FB, Radwan A, Alam M, Rodriguez J (2013) Cooperative strategies for power saving in multi-standard wireless devices. In: Galis A, Gavras A (eds) The future internet. Springer, Berlin, pp 284–296
9. Allot Mobile Trends Report (2014) Measuring the mobile video experience. New insight on video delivery quality in mobile networks
10. Braden R, Clark D, Shenker S (1994) Integrated Services in the internet architecture: an overview. IETF RFC 1633
11. Braden R, Zhang L, Berson S, Herzog S, Jamin S (1997) Resource Reservation Protocol (RSVP) – Version 1 Functional Specification. IETF RFC 2205
12. Manner J, Fu X (2005) Analysis of existing quality-of-service signalling protocols. IETF RFC 4094
13. Changho Y, Harry P (2010) QoS control for NGN: a survey of techniques. Springer J Network Syst Manage 18(4):447–461
14. Blake S, Black D, Carlson M, Davies E, Wang Z, Weiss W (1998) An architecture for differentiated services. IETF RFC 2475
15. Baker F, Iturralde C, Le Faucheur F, Davie B (2001) Aggregation of RSVP for IPv4 and IPv6 reservations. IETF RFC 3175
16. Bless R (2004) Dynamic aggregation of reservations for internet services. Telecommun Syst 26(1):33–52
17. Neto A, Cerqueira E, Curado M, Monteiro E, Mendes P (2008) Scalable resource provisioning for multi-user communications in next generation networks. In: IEEE Global Telecommunications Conference (IEEE GLOBECOM), New Orleans, LA, USA
18. Prior R, Susana S (2007) Scalable Reservation-Based QoS Architecture—SRBQ. In: Encyclopedia of internet technologies and applications, doi: 10.4018/978-1-59140-993-9.ch067
19. Logota E, Campos C, Sargento S, Neto A (2013) Advanced multicast class-based bandwidth over-provisioning. Comput Networks 57(9):2075–2092. doi:10.1016/j.comnet.2013.04.009
20. Castrucci M, Cecchi M, Priscoli FD, Fogliati L, Garino P, Suraci V (2011) Key concepts for the Future Internet architecture. Future Network & Mobile Summit (FutureNetw), Warsaw, Poland
21. Feamster N et al (2004) The case for separating routing from routers. In: Proceedings of ACM SIGCOMM workshop on future directions in network architecture
22. The Open Networking Foundation (2013) OpenFlow Switch Specification, Version 1.4.0 (Wire Protocol 0x05)
23. Logota E, Sargento S, Neto A (2010) Um Método para Controlo Avançado de Sobre-reservas Baseado em Classes de Serviço e Sistema para a sua Execução (A Method and apparatus for Advanced Class-based Bandwidth Over-reservation Control. Patent 105305
24. IETF Mobile Ad Hoc Network (MANET) Working Group. http://datatracker.ietf.org/wg/manet/documents. Accessed 10 July 2014
25. Cover T, Gamal AE (1979) Capacity theorems for the relay channel. IEEE Trans Inform Theory 25(5):572–584

26. Van der Meulen EC (1971) Three-terminal communication channels. Adv Appl Probab 3:120–154
27. Laneman JN, Tse DNC, Wornell W (2004) Cooperative diversity in wireless networks: efficient protocols and outage behavior. IEEE Trans Inform Theory 50(12):3062–3080
28. Laneman JN, Wornell GW (2003) Distributed space-time coded protocols for exploiting cooperative diversity in wireless networks. IEEE Trans Inform Theory 49(10):2415–2425
29. Sendonaris A, Erkip E, Azhang B (2003) User cooperation diversity – part I: system description. IEEE Trans Commun 51(11):1927–1938
30. Sendonaris A, Erkip E, Azhang B (2003) User cooperation diversity – part II: implementation aspects and performance analysis. IEEE Trans Commun 51(11):1939–1948
31. Hunter TE, Nosratinia A (2006) Diversity through coded cooperation. IEEE Trans Wireless Commun 5(2):283–289
32. Janani M, Hedayat A, Huntter TE, Nosratinia A (2004) Coded cooperation in wireless communications: space-time transmission and iterative coding. IEEE Trans Signal Process 52(2):362–371
33. Sadek AK, Su W, Liu KJR (2007) Multinode cooperative communications in wireless networks. IEEE Trans Signal Process 55(1):341–355
34. Boyer J, Falconer DD, Yanikomeroglu H (2004) Multihop diversity in wireless relaying channels. IEEE Trans Commun 52(10):1820–1830
35. Kramer G, Gaspar M, Gupta P (2005) Cooperative strategies and capacity theorems for relay networks. IEEE Trans Inform Theory 51(9):3037–3063
36. Zheng L, Tse DNC (2003) Diversity and multiplexing: a fundamental tradeoff in multiple-antenna channels. IEEE Trans Inform Theory 49(5):1073–1096
37. Nosratinia A, Hunter TE, Hedayat A (2004) Cooperative communication in wireless networks. IEEE Commun Mag 42(10):74–80
38. Nagpal V (2012) Cooperative multiplexing in wireless relay networks. Dissertation, University of California, Berkeley
39. Yijia F, Chao W, Poor HV, Thompson JS (2009) Cooperative multiplexing: toward higher spectral efficiency in multiple-antenna relay networks. IEEE Trans Inform Theory 55 (9):3909–3926
40. Lin Y, Hsu Y (2000) Multihop cellular: a new architecture for wireless communications. In: Proceedings of 19th IEEE conference on INFOCOM, vol. 3, pp 1273–1282
41. Le L, Hossain E (2007) Multihop cellular networks: potential gains, research challenges, and a resource allocation framework. IEEE Commun Mag 45(9):66–73
42. Salem NB, Buttyán L, Hubaux JP, Jakobsson M (2006) Node cooperation in hybrid ad hoc networks. IEEE Trans Mobile Comput 5(4):365–376
43. 3GPP (1999) TR 25.924 version 1.0.0.0: Opportunity Driven Multiple Access
44. 3GPP (2010) TR 36.806 V9.0.0: Relay Architecture for E-UTRA (LTE-Advanced) (Release 9)
45. 3GPP (2012) TS 36.216 V11.0.0: Evolved Universal Terrestrial Radio Access (E-UTRA); Physical layer for relaying operation (Release 11)
46. 3GPP (2014) TR 36.836 V2.1.0: Study on Mobile Relay for Evolved Universal Terrestrial Radio Access (E-UTRA) (Release 12)
47. Iwamura M, Takahashi H, Nagata S (2010) Relay technology in LTE-advanced. NTT Docomo Technical J 12(2):29–36
48. Aggélou GN, Tafazolli R (2001) On the relaying capability of next-generation GSM cellular networks. IEEE Personal Commun 8(1):40–47
49. Pabst R et al (2004) Relay-based deployment concepts for wireless and mobile broadband radio. IEEE Commun Mag 45(9):80–89
50. FP7 EU Project (2014) C2POWER: Cognitive radio and cooperative strategies for power saving in multi-standard wireless devices. http://www.ict-c2power.eu. Accessed 9 July 2014

51. FP7 EU Project (2014) GREENT: Green terminals for next generation wireless systems. http://greent.av.it.pt. Accessed 9 July 2014
52. Radwan A, Rodriguez J (2012) Energy saving in multi-standard mobile terminals through short-range cooperation. EURASIP J Wireless Commun Networking 2012(159):1–15
53. Saghezchi FB, Radwan A, Rodriguez J (2013) Energy efficiency performance of wiFi/WiMedia relaying in hybrid ad-hoc networks. In: Proceedings of IEEE 3rd International Conference on Communications and Information Technology (ICCIT), Beirut, pp 285–289
54. MacKenzie AB, DaSilva LA (2006) Game theory for wireless engineers. Morgan and Claypool Publishers, California
55. Felegyhazi M, Hubaux J (2006) Game theory in wireless networks: a tutorial. Technical Report LCA- REPORT-2006-002, EPFL
56. Saad W, Han Z, Debbah M, Hjørungnes A, Basar T (2009) Coalitional game theory for communication networks. IEEE Signal Process Mag 26(5):77–97
57. Saghezchi FB, Nascimento A, Albano M, Radwan A, Rodriguez J (2011) A novel relay selection game in cooperative wireless networks based on combinatorial optimizations. In: Proceedings of IEEE 73rd Vehicular Technology Conference (VTC Spring), Budapest
58. Saghezchi FB, Radwan A, Nascimento A, Rodriguez J (2012) An incentive mechanism based on coalitional game for fair cooperation of mobile users in HANETs. In: Proceedings of IEEE 17th International Workshop on Computer Aided Modeling and Design of Communication Links and Networks (CAMAD), Barcelona, pp 378–382
59. Saghezchi FB, Radwan A, Rodriguez J, Dagiuklas T (2013) Coalition formation game toward green mobile terminals in heterogeneous wireless networks. IEEE Wireless Commun Mag 20 (5):85–91. doi:10.1109/MWC.2013.6664478
60. Srivastava V et al (2005) Using game theory to analyze wireless ad hoc networks. IEEE Commun Surv Tutorials 7(4):46–56
61. Félegyázi M, Hubaux J, Buttyán L (2006) Nash equilibrium of packet forwarding strategies in wireless ad hoc networks. IEEE Trans Mobile Comput 5(5):463–476
62. Yang J, Klein AG, Brown DR III (2009) Natural cooperation in wireless networks. IEEE Signal Process Mag 26(5):98–106
63. Srinivasan V, Nuggehalli P, Chiasserini CF, Rao RR (2005) An analytical approach to the study of cooperation in wireless ad hoc networks. IEEE Trans Wireless Commun 4 (2):722–733
64. Buttyán L, Hubaux, JP (2001) Nuglets: a virtual currency to stimulate cooperation in self organized mobile ad hoc networks. Technical report, no. DSC/2001
65. Jakobsson M, Hubaux JP, Buttyán L (2003) A micro-payment scheme encouraging collaboration in multi-hop cellular networks. Financial cryptography. Springer, Berlin, pp 15–33
66. Salem NB, Levente B, Hubaux JP, Jakobsson M (2003) A charging and rewarding scheme for packet forwarding in multi-hop cellular networks. In: Proceedings of MOBIHOC'03, Maryland, USA
67. Zhong S, Chen J, Yang YR (2003) Sprite: a simple, cheat- proof, credit-based system for mobile ad hoc networks. In: IEEE INFOCOM'03, vol. 3, San Francisco, pp 1987–1997
68. Buchegger S, Boudec JL (2002) Performance analysis of the CONFIDANT protocol. In: Proceedings of the 3rd ACM International Symposium on Mobile Ad Hoc Networking & Computing, Lausanne, pp 226–236
69. Michiardi P, Molva R (2002) CORE: a collaborative reputation mechanism to enforce node cooperation in mobile ad hoc networks. In: Proceedings of IFIP communication and multi-media security conference
70. Bansal S, Baker M (2003) Observation-based cooperation enforcement in ad hoc networks. http://arxiv.org/pdf/cs/0307012v2. Accessed 9 July 2014
71. He Q, Wu D, Khosla P (2004) SORI: A Secure And Objective Reputation-based Incentive Scheme for Ad Hoc Networks. In: IEEE wireless communications and networking conference, pp 825–830

72. Rebahi Y, Mujica V, Simons C, Sisalem D (2005) SAFE: Securing packet forwarding in Ad Hoc networks. In: Proceeding of the 5th Workshop on Applications and Services in Wireless Networks (ASWN), Paris
73. Jaramillo JJ, Srikant R (2007) DARWIN: distributed and adaptive reputation mechanism for wireless ad-hoc networks. In: Proceedings of the 13th Annual ACM International Conference on Mobile Computing and Networking (MOBICOM'07), Montréal
74. Saghezchi FB, Radwan A, Rodriguez J, Taha AM (2014) Coalitional relay selection game to extend battery lifetime of multi-standard mobile terminals. In: IEEE International Conference on Communications (ICC), Sydney, pp 508–513
75. Saghezchi FB, Radwan A, Rodriguez J (2014) A coalitional game-theoretic approach to isolate selfish nodes in multihop cellular network. In: The 9th IEEE Symposium on Computers and Communications, (ISCC), Madeira
76. ITU-T Recommendation (2004) Y.2001: General overview of NGN
77. http://www.cisco.com/en/US/docs/solutions/Enterprise/WAN_and_MAN/QoS_SRND/QoSIntro.html. Accessed 30 June 2014
78. Sofia R (2004) SICAP, a shared-segment inter-domain control aggregation protocol. Dissertation, Universidade de Lisboa, Portugal
79. Logota E, Neto A, Sargento S (2010) A new strategy for efficient decentralized network control. In: IEEE Global Telecommunications Conference, (IEEE GLOBECOM)
80. Golestani S-J (1994) A self-clocked fair queueing scheme for broadband applications. In: INFOCOM. Networking for Global Communications, Toronto, pp 12–16
81. Evariste L, Campos C, Sargento S, Neto A (2013) Scalable resource and admission management in class-based networks. In: IEEE International Conference on Communications 2013 (ICC'13) – 3rd IEEE International Workshop on Smart Communication Protocols and Algorithms (SCPA 2013) – ('ICC'13–IEEE ICC'13–Workshop SCPA'), Budapest, Hungary
82. The Zettabyte Era — Trends and Analysis. http://www.cisco.com/en/US/solutions/collateral/ns341/ns525/ns537/ns705/ns827/VNI_Hyperconnectivity_WP.html. Accessed 30 June 2014
83. ITU-T Study Group 12 (2008) Performance monitoring points for IPTV. ITU-T Recommendation G.1081
84. Lima SR, Carvalho P (2011) Enabling self-adaptive QoE/QoS control. In: Proceedings of IEEE 36th Conference on Local Computer Networks (LCN), Bonn, pp 239–242
85. Salehin KM, Rojas-Cessa R (2010) Combined methodology for measurement of available bandwidth and link capacity in wired packet networks. IET Commun 4(2):240–252
86. Moi J (1998) OSPF version 2. IETF RFC2328
87. Pan P, Hahne E, Schulzrinne H (2000) BGRP: a tree-based aggregation protocol for inter-domain reservations. J Commun Networks 2(2):157–167
88. Pinto P, Santos A, Amaral P, Bernardo L (2007) SIDSP: Simple inter-domain QoS signaling protocol. In: Proceedings of IEEE Military Communications Conference, Orlando, FL, USA
89. Bless R (2002) Dynamic aggregation of reservations for internet services. In: Proceedings of 10th Int. Conf. on Telecommunication. Systems – Modeling and Analysis (ICTSM'10), vol 1, pp 26–38
90. Sofia R, Guerin R, Veiga P (2003) SICAP, a shared-segment inter-domain control aggregation protocol. In: Proceedings of Int. Conf. in High Performance Switching and Routing, Turin, Italy
91. The Network Simulator – ns-2.31. http://www.isi.edu/nsnam/ns/. Accessed 7 July 2014
92. Babiarz J, Chan K, Baker F (2006) Configuration Guidelines for DiffServ Service Classes. IETF RFC 4594
93. Demers A, Keshav S, Shenker S (1989) Analysis and simulation of a fair queueing algorithm. In: ACM SIGCOMM'89, vol. 19, pp 1–12
94. Evariste L, Hugo M, Jonathan R (2013) A cross-layer resource over-provisioning architecture for P2P networks. In: Proceedings of 18th international conference on digital signal processing – special session on 3D immersive & interactive multimedia over the future internet, Santorini, Greece

95. Katz D (1997) IP Router Alert Option. IETF RFC 2113, Feb 1997
96. Hancock R, Karagiannis G, Loughney J, Van den Bosch S (2005) Next Steps in Signalling (NSIS): Framework. IETF RFC 4080
97. Manner J, Karagiannis G, McDonald A (2008) NSLP for quality-of-service signaling. Draft-ietf-nsis-qos-nslp-16 (work in progress)
98. Vasseur J-P, Ali Z, Sivabalan S (2006) Definition of a Record Route Object (RRO) Node-Id Sub-Object. IETF RFC 4561
99. Rekhter Y, Li T, Hares S (2006) A Border Gateway Protocol 4 (BGP-4). IETF RFC 4271
100. Rosenberg J et al (2002) SIP: Session Initiation Protocol. IETF RFC 3261

Chapter 9
Real-Time 3D QoE Evaluation of Novel 3D Media

Chaminda T.E.R. Hewage, Maria G. Martini, Harsha D. Appuhami, and Christos Politis

Abstract Recent wireless networks enable the transmission of high bandwidth multimedia data, including advanced 3D video applications. Such wireless multimedia systems should be designed with the purpose of maximizing the quality perceived by the users. For instance, quality parameters can be measured at the receiver-side and fed back to the transmitter for system optimization. Measuring 3D video quality is a challenge due to a number of perceptual attributes associated with 3D video viewing (e.g., image quality, depth perception, naturalness). Subjective as well as objective metrics have been developed to measure 3D video quality against different artifacts. However most of these metrics are *Full-Reference* (*FR*) quality metrics and require the original 3D video sequence to measure the quality at the receiver-end. Therefore, these are not a viable solution for system monitoring/update "on the fly." This chapter presents a *Near No-Reference* (*NR*) quality metric for color plus depth 3D video compression and transmission using the extracted edge information of color images and depth maps. This work is motivated by the fact that the edges/contours of the depth map and of the corresponding color image can represent different depth levels and identify image objects/boundaries of the corresponding color image and hence can be used in quality evaluation. The performance of the proposed method is evaluated for different compression ratios and network conditions. The results obtained match well those achieved with its counterpart *FR* quality metric and with subjective tests, with only a few bytes of overhead for the original 3D image sequence as side-information.

This work was supported in part by the EU FP7 Programme (CONCERTO project).

C.T.E.R. Hewage (✉) • M.G. Martini • H.D. Appuhami • C. Politis
Kingston University London, Kingston Upon Thames, KT1 2EE, UK
e-mail: c.hewage@kingston.ac.uk

© Springer Science+Business Media New York 2015
A. Kondoz, T. Dagiuklas (eds.), *Novel 3D Media Technologies*,
DOI 10.1007/978-1-4939-2026-6_9

163

9.1 Introduction

The recently standardized wireless systems, including WIMAX, 3GPP LTE/LTE-Advanced (LTE-A), the latest 802.11 standards, and advanced short range wireless communication systems, enable the transmission of high bandwidth multimedia data such as immersive 3D video applications. For such applications in particular, the target of the system design should be the maximization of the final quality perceived by the user, or quality of experience (QoE), rather than only of the performance of the network in terms of "classical" quality of service (QoS) parameters such as throughput and delay. Moreover, the overall enjoyment or annoyance of 3D video streaming applications or services is influenced by several aspects such as human factors (e.g., demographic and socio-economic background), system factors (e.g., content- and network-related influences), and contextual factors (e.g., duration, time of the day, and frequency of use). The overall experience can be analyzed and measured by QoE-related parameters which quantify the user's overall satisfaction about a service [1, 2]. QoS-related measurements only measure performance aspects of a physical system, with main focus on telecommunications services. Measuring QoS parameters is straightforward since objective, explicit technological methods can be used, whereas measuring and understanding QoE requires a multidisciplinary and multi-technological approach. The added dimension of depth in 3D viewing influences several perceptual attributes such as overall image quality, depth perception, naturalness, presence, visual comfort, etc. For instance, an increased binocular disparity enhances the depth perception of viewers, although in extreme cases this can lead to eye fatigue as well. Therefore, the overall enjoyment of the 3D application could be hindered by the eye strain experienced by the end user. The influence of these attributes on the overall experience of 3D video streaming users is yet to be investigated.

3D video services delivered through wireless and mobile channels face a number of challenges due to the need to handle a large amount of data and to the possible limitations due to the characteristics of the transmission channel and of the device. This can result in perceivable impairments originated in the different steps of the communication system, from content production to display techniques, and influence the user's perception of quality. For instance channel congestion and errors at the physical layer may result in packet losses and delay, whereas compression techniques introduce compression artifacts. Such impairments could be perceived by the end user and result to a different extent in the degradation of the quality of the rendered 3D video. For some 3D image formats such as color plus depth 3D video representations cause adverse effects when they are subjected to impairments since depth map component is used to project the corresponding color image into the 3D space.

The different processing steps along the end-to-end 3D video chain introduce image artifacts which may affect 3D perception and the overall experience of viewers [3]. Even though much attention has been paid into analyzing and mitigating the effects of 3D image/video capture, processing, rendering and display techniques, the effects of artifacts introduced by the transmission system have not received much attention compared to the 2D image/video counterpart. Some of these artifacts influence the overall image quality, for instance blurriness, luminance, and contrast levels, similar as in 2D image/video. The effect of transmission over band-limited and unreliable communication channels (such as wireless channels) can be much worse for 3D video than for 2D video, due to the presence of two channels (i.e., stereoscopic 3D video) that can be impaired in a different way; as a consequence the 3D reconstruction in the human visual system may be affected. Some networks introduce factors directly related to temporal domain de-synchronization issues. For instance delay in one view could lead to temporal de-synchronization and this can lead to reduced comfort in 3D viewing.

The methods employed to mitigate these artifacts (e.g., error concealment) need to be carefully designed to suit 3D video applications. The simple application of 2D image/video methods would not work effectively for 3D image/video, as discussed in [4] for different error concealment algorithms for 3D video transmission errors. In [4], it is observed that in some cases switching back to the 2D video mode is preferred to applying 2D error concealment methods separately for left and right views to recover missing image information during transmission. There could be added implications introduced by these artifacts into our HVS. Therefore artifacts caused as a result of 3D video streaming can be clearly appreciated only by understanding how our HVS perceives different 3D video artifacts.

The HVS is capable of aligning and fusing two slightly different views fed into the left and right eyes and hence of perceiving the depth of the scene. Both binocular and monocular cues assist our HVS to perceive different depth planes of image objects [5]. Binocular disparity is the major cue used by the HVS to identify the relative depth of objects. Other monocular depth cues include perspective, occlusions, motion parallax, etc. ([5]). During 3D video streaming, one view or both views could be badly affected by channel impairments (e.g., bit errors and packet losses caused by adverse channel conditions, delay, jitter). For instance, frame freezing mechanisms employed to tackle missing frames caused by transmission errors or delay could lead to temporal de-synchronization where one eye sees delayed content compared to the other eye. There are two implications associated to the case where one view is affected by transmission impairments;

- Binocular suppression
- Binocular rivalry

Our HVS is still capable to align and fuse stereoscopic content if one view is affected by artifacts due to compression, transmission, and rendering. Binocular suppression theory suggests that in these situations the overall perception is usually driven by the quality of the best view (i.e., left or right view), at least if the quality of the worst view is above a threshold value. However this capability is limited and studies show that additional cognitive load is necessary to fuse these views [6]. Increased cognitive load leads to visual fatigue and eye strain and prevents users from watching 3D content for a long time. This directly affects user perception and QoE. If one of the views is extremely altered by the transmission system, the HVS will not be able to fuse the affected views, and this causes binocular rivalry. This has detrimental effects on the final QoE perceived by the end user. Recent studies on 3D video transmission [4] have found that binocular rivalry is influencing the overall perception and this effect prevails over the effect of binocular suppression. To avoid the detrimental effect of binocular rivalry, the transmission system could be designed appropriately taking this issue into account. For instance, the transmission system parameters can be updated "on the fly" to obtain 3D views with minimum distortions, according to the feedback on the measure of 3D video quality at the receiver-side. In case of low quality due to different errors in the two views, if the received quality of one of the views is significantly low, the transmission system could be informed to allocate more resources to the worse view or to increase the error protection level for that 3D video channel to mitigate the quality loss in subsequent frames. This increases the opportunity to fuse the 3D video content more effectively and improve the final QoE of users. The measured image quality at the receiver-side can be used as feedback information to update system parameters "on the fly" in a "QoE-aware" system design approach as discussed in [2]. However, measuring 3D video quality is a challenge mainly due to the complex nature of 3D video quality [7] and the necessity of sending a copy of the original 3D image sequence for measuring the quality with FR methods.

Immersive video quality evaluation is a hot topic among researchers and developers at present, due to its complex nature and to the unavailability of an accurate objective quality metric for 3D video. 3D perception can be associated with several perceptual attributes such as "overall image quality," "depth perception," "naturalness," "presence," "comfort," etc. A detailed analysis is necessary to study how these 3D percepts influence the overall perceived 3D image quality in general. For instance, the study presented in [8] concludes that excessive disparities between left and right view can cause eye strain and therefore degrade the perceived image quality. Mostly, appreciation-oriented psychophysical experiments are conducted to measure and quantify 3D perceptual attributes. A few standards also define subjective quality evaluation procedures for both 2D and 3D video (e.g., [9, 10]). However, these procedures are not competent enough to measure 3D QoE parameters and show several limitations; for instance these are not able to measure the combined effect of different perceptual attributes. Subjective quality evaluation under different system parameter changes have been reported in a number of studies [8, 11, 12]. However, these studies are limited to certain types of image artifacts (e.g., compression artifacts) and have limited usage in practical applications. On the other hand, subjective quality evaluation requires time, effort, controlled test environments, money, human observers, etc. and cannot be deployed in a live environment where quality is measured "on the fly."

Objective quality evaluation methods for 3D video are also emerging to provide accurate results in comparison to the quality ratings achieved with subjective tests [13, 14]. However, the performance of these metrics is most of the time an approximation to that of subjective quality assessments. Our recent studies have also found out that there is a high correlation between subjective ratings and individual objective quality ratings of 3D video components (e.g., average PSNR and SSIM of left and right video or color and depth video) [15–17]. For instance, "depth perception" is highly correlated to the average PSNR of the rendered left and right image sequences [15]. This means that we could use individual objective quality measures of different 3D video components to predict the true user perception in place of subjective quality evaluation, through a suitable approximation derived based on correlation analysis. However, with some 3D source representations such as color and depth map 3D image format, it may be difficult to derive a direct relationship between objective measures and subjective quality ratings. For instance, objective quality of depth map may have a very weak correlation on its own for overall subjective quality, because depth map is used for projecting the corresponding color image into 3D coordinates and it is not directly viewed by the end users. All the methods described above are FR methods and need the original 3D image sequence to measure the quality, hence they are less useful in real-time transmission applications. The solution is to use Reduced-Reference (RR) or No-Reference (NR) metrics which need fewer or zero bits respectively to transmit the side-information for the original image sequence. RR and NR quality metrics are proposed in the literature for 2D video [18–21], but only a few are reported so far for 3D video [22]. For instance, [22] proposed a NR quality metric for 3D images based on left and right stereoscopic images. The proposed Near NR quality assessment in this chapter considers instead stereoscopic image sequences based on color plus depth 3D video format. Due to recent advances in free viewpoint 3D video delivery with multi-color and multi-depth map 3D format [23], quality metrics as proposed in this chapter could be effectively used in emerging applications.

Color plus depth 3D video enables us to render virtual left and right views based on a 3D image warping method commonly known as Depth Image Based Rendering (DIBR) [24]. This representation consists of two images per video frame, one representing color information, similar to 2D video, the other representing depth information. The edge information of both color and depth map images can represent different depth levels and object boundaries (see Fig. 9.1). The major boundaries of image objects of both color and depth are coincident. This information can be exploited to quantify irregularities of color and depth map sequences. This chapter therefore proposes a Near NR quality evaluation approach for color plus depth format 3D video based on edge information characteristics. Edge information extracted from images has been employed in the past in measuring 2D image quality (e.g., [25, 26]). In our previous work, a RR metric utilizing gradient information was proposed for depth maps associated with color plus depth based 3D video [27, 28]. It shows comparable results to FR methods, but limited only for measuring depth map quality. Furthermore, the overhead for side-information, although limited, was of the order of kbps. In this chapter, we propose a global quality evaluation method for color plus depth 3D video based on edge extraction, requiring only a few bytes per second for the transmission of side-information (Near NR).

This chapter is organized as follows: Sect. 9.2 reports the proposed Near NR quality metric for color plus depth 3D video transmission. The experimental setup, results, and discussion are presented in Sect. 9.3. Section 9.4 concludes the chapter.

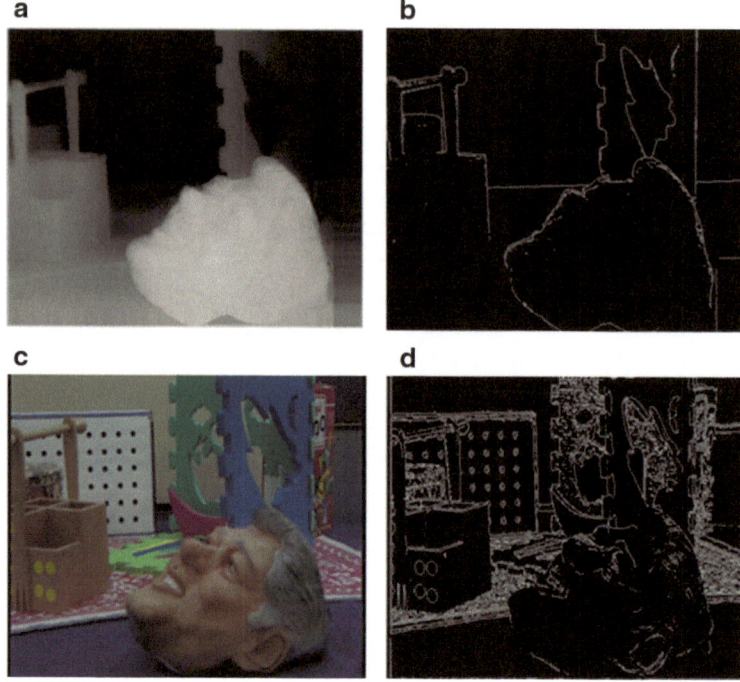

Fig. 9.1 The *Orbi* sequence (**a**) depth map, (**b**) extracted binary edge mask of the depth map using *Sobel* filtering, (**c**) corresponding color image, and (**d**) extracted binary edge mask of the color image using *Sobel* filtering

9.2 Proposed Real-Time Quality Evaluation Strategy

The color image and the corresponding depth map share common object boundaries even though individual pixel values are different (see Fig. 9.1). The gradient/edge information of these images can be used to detect object boundaries/contours.

The edge information of the depth map represents depth map object boundaries and different depth levels (see Fig. 9.1b). Depth discontinuities occur around these edges; hence, their preservation is vital in 3D video rendering based on the DIBR method [29], since any imperfections around these edges will severely affect the quality of the rendered left and right views. The effectiveness of using edge information for measuring the depth map quality is shown in [27, 28]. Similarly to the depth edge characteristics, gradient information of the color image sequence also identifies object contours and boundaries (see Fig. 9.1d) and is also used in color image quality evaluation [25, 26]. Figures 9.2 and 9.3 demonstrate the effectiveness of using gradient information to represent artifacts of the images due to compression and packet losses occurring during transmission. In these figures, both 3D sequences are encoded using the Quantization Parameter $(QP) = 30$ and corrupted with 20 % *PLR* [30]. This shows that edge information can be utilized to represent the structural degradation of color plus depth 3D images. Subsequently, this information together with additional data (e.g., *luminance*, *contrast*) can be used to evaluate the quality of corrupted 3D video sequences. The changes in edge map will not count for pixel level changes of the original and processed color and depth map sequences. Therefore supplementary data is required to measure pixel domain changes of these sequences. The edge information of color and depth map image sequences, together with some additional information such as *luminance* and *contrast*, is compared to obtain quality ratings for 3D video in the proposed method.

The schematic diagram of the proposed *Near NR* quality evaluation method is shown in Fig. 9.4. At the receiver-side, edge information is extracted from both the color and depth map images. In addition to this *structural* measure based on edge information, *luminance* and *contrast* information from the original and processed

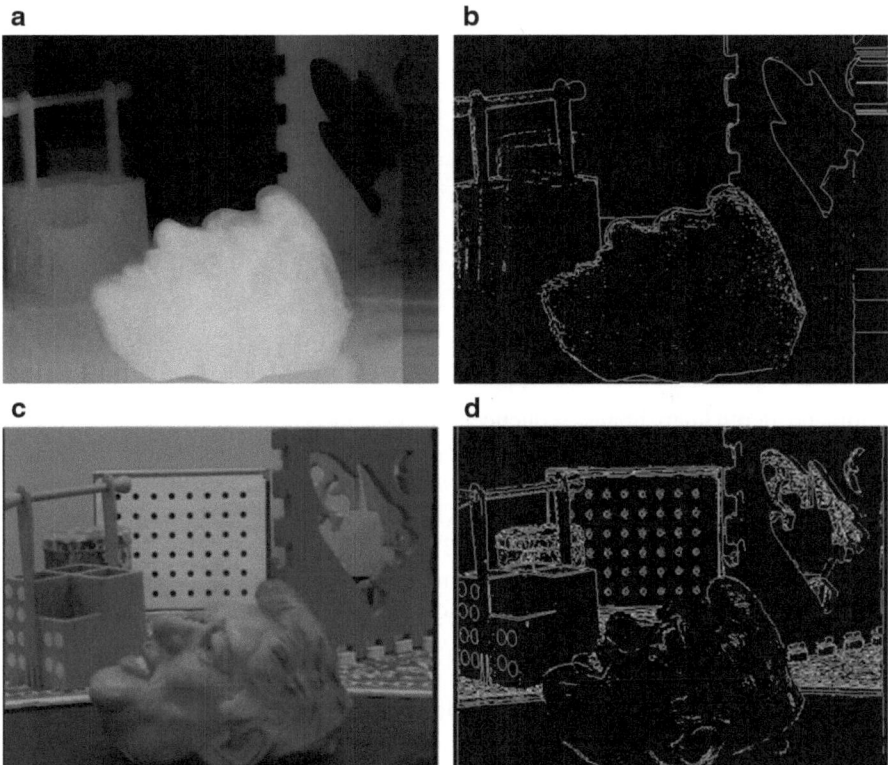

Fig. 9.2 Graphical illustration of the proposed quality metric using the *Orbi* sequence (50th frame) coded using *QP* 30 and transmitted over a network with *PLR* 20 %. (**a**) original depth map; (**b**) extracted edge information from the original depth map; (**c**) original color image; (**d**) extracted edge information from the original color image; (**e**) processed depth map; (**f**) extracted edge information from the processed depth map; (**g**) processed color image, and (**h**) extracted edge information from the processed color image

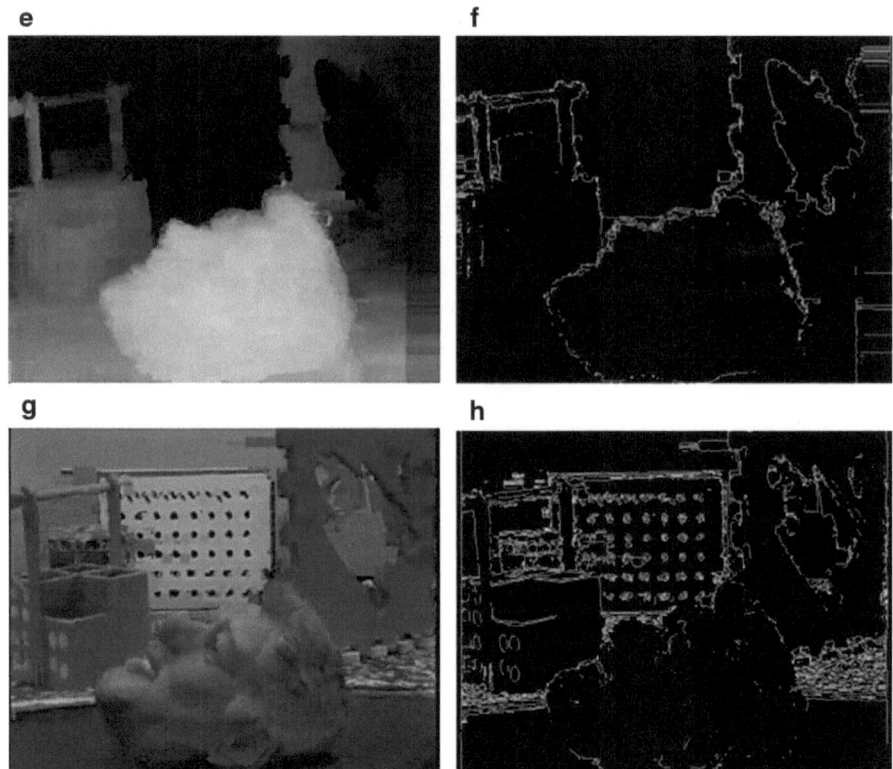

Fig. 9.2 (continued)

color and depth map image sequences are considered to obtain *luminance* and *contrast* measures of the metric. *Luminance* and *contrast* information calculated at the sender side only occupies a few bytes per frame. This justifies that we could name the proposed method as a *Near NR* metric rather than a *RR* method. The structural similarity (SSIM) metric in [31] is modified to measure the effect of *luminance*, *contrast*, and *structural* information in the proposed method. Equations (9.1), (9.2), (9.3), (9.4), and (9.5) show how the SSIM calculation is performed.

Structural similarity (SSIM) index between signal x and y:

$$SSIM(x,y) = [l(x,y)]^{\alpha}.[c(x,y)]^{\beta}.[s(x,y)]^{\gamma} \tag{9.1}$$

where $SSIM(x, y)$ represents the similarity map, $\alpha > 0$, $\beta > 0$, and $\gamma > 0$ and $l(x, y)$, $c(x, y)$, and $s(x, y)$ represent *luminance*, *contrast*, and *structural* comparisons respectively. $l(x, y)$, $c(x, y)$, and $s(x, y)$ components are given by (9.2), (9.3), and (9.4) respectively.

$$l(x, y) = \frac{2\mu_x\mu_y + C_1}{\mu_x^2 + \mu_y^2 + C_1} \tag{9.2}$$

$$c(x, y) = \frac{2\sigma_x\sigma_y + C_2}{\sigma_x^2 + \sigma_y^2 + C_2} \tag{9.3}$$

$$S(x, y) = \frac{\sigma_{xy} + C_3}{\sigma_x\sigma_y + C_3} \tag{9.4}$$

where x and y are block vectors (i.e., local 8×8 square window) of images X and Y respectively, μ_x and μ_y are the mean of vector x and y respectively, σ_x and σ_y are the standard deviation of vector x and y respectively, σ_{xy} is the covariance of vector x and y, and C_1, C_2, and C_3 are small constants to avoid the denominators being zero. Mean and standard deviation are calculated in block based (each block is 8×8). The *luminance* comparison mainly provides a rating for the average

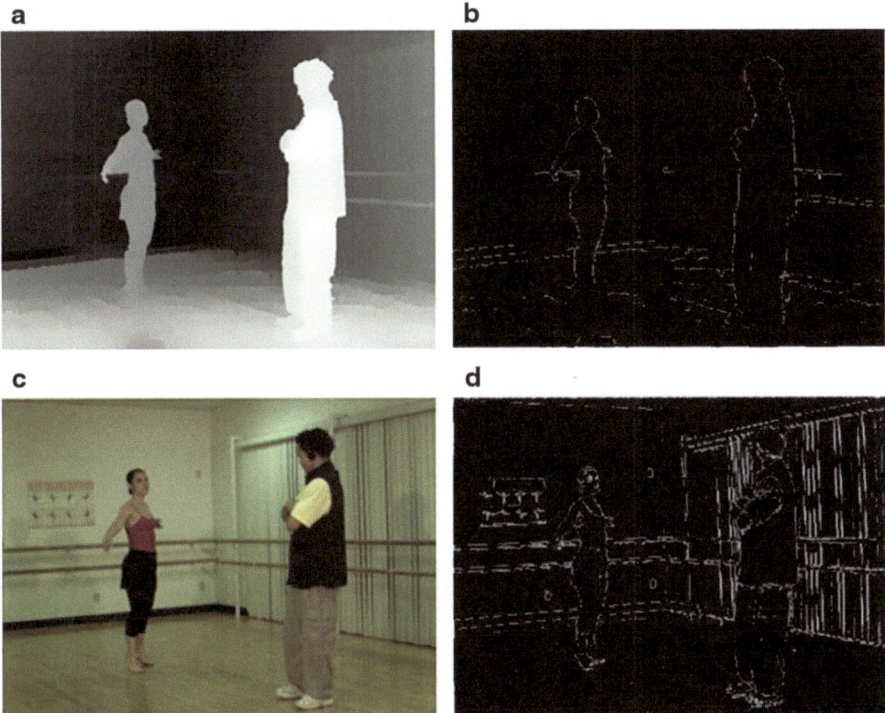

Fig. 9.3 Graphical illustration of the proposed quality metric using the *Ballet* sequence (30th frame) coded using *QP* 30 and transmitted over a network with *PLR* 20 %. (**a**) original depth map; (**b**) extracted edge information from the original depth map; (**c**) original color image; (**d**) extracted edge information from the original color image; (**e**) processed depth map; (**f**) extracted edge information from the processed depth map; (**g**) processed color image, and (**h**) extracted edge information from the processed color image

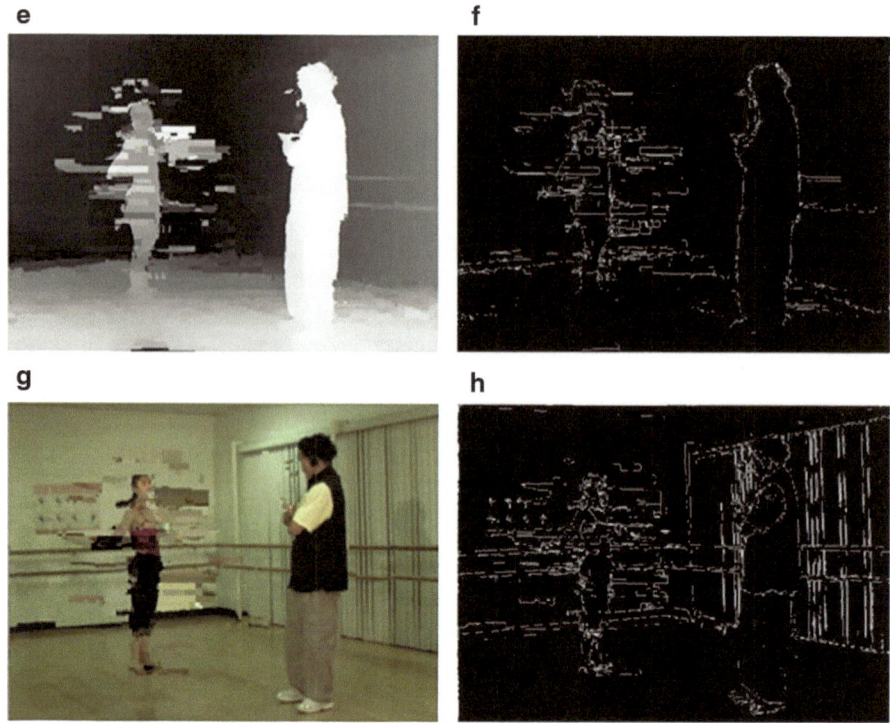

Fig. 9.3 (continued)

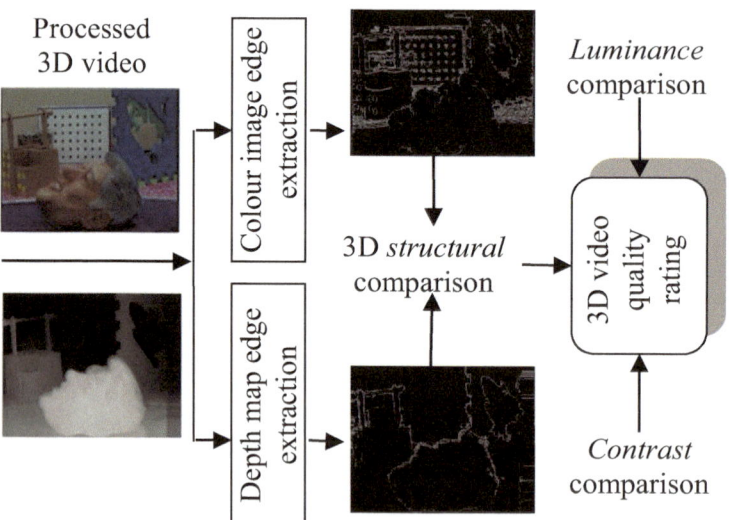

Fig. 9.4 Block diagram of the proposed *Near NR* method

intensity of the image (using the mean value) whereas *contrast* comparison provides an indication about the variation of pixel intensity of the image frame (using the standard deviation of individual images). *Structural* comparison provides an indication about structural degradation compared to the original image by calculating the covariance between the original and processed images. Mean SSIM (MSSIM) index for overall image quality evaluation is defined as;

$$MSSIM(X, Y) = \frac{1}{M} \sum_{j=1}^{M} SSIM(x_j, y_j) \tag{9.5}$$

where X and Y are the reference and the distorted images, respectively; x_j and y_j are the image contents at the jth local window; and M is the number of local windows in the image. MSSIM represents the mean SSIM and it is a measure of the structure similarity between images X and Y.

In the proposed method, *luminance* and *contrast* comparisons are performed based on image statistics (*mean* and *standard deviation* of individual images) generated at the sender and receiver ends (see [31] for further details about calculating *luminance* and *contrast* comparisons). The *structural* comparison is performed between the generated edge information of the processed color and corresponding depth map images. If X' and Y' represent the edge/gradient maps of the received color and the received depth map image respectively, the edge/gradient based structural comparison $s_e(x,y)$ can be defined as;

$$S_e(x, y) = \frac{\sigma_{x'y'} + C}{\sigma_{x'}\sigma_{y'} + C} \tag{9.6}$$

where x' and y' are block vectors of X' and Y' respectively, $\sigma_{x'}$ and $\sigma_{y'}$ are the standard deviation of vector x' and y' respectively, $\sigma_{x'y'}$ is the covariance of vector x' and y', and C is a small constant to avoid the denominator being zero. Then the edge based structural similarity metric (E-SSIM) can be defined as follows;

$$E\text{-}SSIM = [l(x, y)]^\alpha \cdot [c(x, y)]^\beta \cdot [s_e(x, y)]^\gamma \tag{9.7}$$

Then the mean image quality is calculated as;

$$M[E\text{-}SSIM(X, Y)] = \frac{1}{M} \sum_{j=1}^{M} E\text{-}SSIM(x_j, y_j) \tag{9.8}$$

A similar SSIM metric has been proposed for 2D video in [32]. However, in this case the edge information of the original and processed images has been used for structural comparison.

The *structural* comparison term (i.e., $s_e(x,y)$) is common for both color and depth map quality evaluation and only the terms associated with *luminance* and *contrast* comparisons are calculated using individual color and depth map images. The Structural SIMilarity maps for color and depth map sequences (i.e., E-SSIM$_C$ and E-SSIM$_D$) are therefore defined by (9.9) and (9.10) respectively.

$$\text{E-SSIM}_C = [l_C(x,y)]^\alpha \cdot [c_C(x,y)]^\beta \cdot [s_e(x,y)]^\gamma \tag{9.9}$$

$$\text{E-SSIM}_D = [l_D(x,y)]^\alpha \cdot [c_D(x,y)]^\beta \cdot [s_e(x,y)]^\gamma \tag{9.10}$$

where $l_C(x,y)$ and $l_D(x,y)$ are *luminance* comparisons of color and depth map sequences respectively, $C_C(x,y)$ and $C_D(x,y)$ are contrast comparisons of color and depth map sequences respectively, and $S_e(x,y)$ is the *structural* comparison between the edge/gradient maps of the received color and depth map sequences.

Since the proposed method heavily depends on the accuracy of edge detection, a suitable edge detection scheme has to be utilized [33]. In this work, the *Sobel* operator is selected to obtain edge information (i.e., the binary edge mask) due to its simplicity and efficiency [34]: *Sobel* filtering is typically used when real time operation is needed [35]. This is important in scenario where "on-the-fly" system adaptation is performed based on video quality feedback.

In the proposed method, initially the horizontal and vertical edges (i.e., $E_{\text{Horizontal}}(x, y)$ and $E_{\text{Vertical}}(x, y)$) are calculated by applying the *Sobel* filter in each image dimension. Then the gradient image ($G(x, y)$) is calculated as follows;

$$G(x,y) = \sqrt{\left|E_{\text{Horizontal}}(x,y)\right|^2 + \left|E_{\text{Vertical}}(x,y)\right|^2} \tag{9.11}$$

The extracted edge/gradient information ($G(x, y)$) from the processed/received color and depth maps images is employed to quantify the *structural* degradation of the image.

The amount of edges detected in the sequence depends on a number of factors, including the edge detection threshold used and the amount of compression (*QP* values being used), besides the characteristics of the sequence. For instance, when compressed with higher *QP* values, images are smoothed (due to the removal of high frequencies) and *blockiness* can occur when using *DCT* based image encoders, hence detection of accurate edges can be difficult. Therefore, the *structural* comparison performed with the proposed method is sensitive to the compression level being used. This may tend to get saturated *structural* degradation ratings at higher *QP* values.

Even though the gradient/edge information of color and depth map images provides a significant indication of the structural degradation of the processed images, due to the abstract level of information being used with the proposed method, the quality ratings may slightly vary compared to the *FR* SSIM method (the *FR* method uses the complete, original image sequence as the reference).

In order to minimize the difference between the proposed method and *FR* quality evaluation methods, relationships can be derived between them based on experimental findings. For instance, the measured quality with the proposed method (i.e., structural similarity maps E-SSIM$_C$ and E-SSIM$_D$) can be approximated based on

FR quality ratings of the individual color (SSIM$_C$) and depth map (SSIM$_D$) images. Equations (9.12) and (9.13) define the relationships between the proposed and *FR* methods.

$$\text{SSIM}_C \approx f_{\text{Colour}}(\text{E-SSIM}_C) \qquad (9.12)$$

$$\text{SSIM}_D \approx f_{\text{Depth}}(\text{E-SSIM}_D) \qquad (9.13)$$

where SSIM$_C$ and SSIM$_D$ are *FR* SSIM maps of color and depth map images respectively.

Even though the proposed quality estimation obtained with Eqs. (9.12) and (9.13) provides *SSIM* ratings for color and depth map sequences separately, it may not provide direct relationship between these measures and subjective 3D video quality results. This is mainly due to the nature and usage of color and depth map sequences in 3D view rendering (i.e., Depth map is not directly viewed by the users and only used for 3D video rendering.) However, the existing relationship between these objective measures and subjective quality ratings (e.g., overall 3D image quality, depth perception) are studied in our previous research [15–17]. As elaborated in Section 9.1, these studies show a strong correlation between subjective and objective quality ratings (including the *SSIM* metric) for a range of compression rates and *PLR*s. In this chapter also we provide experimental results to show the degree of correlation between individual objective quality measures and true 3D subjective quality ratings obtained after series of subjective quality evaluation tests.

9.3 Experimental Setup, Results, and Discussion

In order to evaluate the performance of the proposed *NR* quality metric for color plus depth 3D video, experiments are performed for different *PLR*s (i.e., 0, 3, 5, 10, and 20 %) and compression levels (i.e., with different *QP* values). The *Orbi*, *Interview*, *Ballet*, and *Breakdance* 3D video test sequences are encoded using the H.264/AVC video coding standard (*JM* reference software Version 16.0). Ten-second long sequences (i.e., 250 frames from *Orbi* and *Interview* sequences-25 fps and 150 frames from *Ballet* and *Breakdance* sequences-15 fps) are encoded with *IPPPIPPP*... format, using *QP* values 1, 5, 10, 15, 20, 25, 30, 35, 40, 45, and 50. An I frame is encoded by every one second. Slices (one row of *MB*s = one slice) are also introduced in order to make the decoding process more robust to errors.

The transmission of the encoded bit-stream is simulated over an IP core network and over an LTE-like wireless system. In the first case, transmission is simulated by using IP error patterns generated for Internet experiments [30]. In order to obtain average results, random starting positions are used for the error pattern files. The corrupted bit-streams with different *PLR*s are later decoded using *JM* reference software decoder.

In the second case, transmission over a wireless channel is simulated considering an OFDMA based transmission system (such as the recent WiMAX and 3GPP LTE systems). All the layers of the OSI protocol stack are realistically simulated, from packetization to transmission over the network through RTP/UDP/IP and over the MAC/Physical layer. The wireless channel is simulated through a Rayleigh block fading channel in time and frequency, with block duration of 0.1 s and channel coherence bandwidth of 1.25 MHz. In addition to fast fading, log-normal block fading is taken into account, in order to model slowly varying shadowing effects ($\sigma[dB] = 8$ dB and block duration of 8 s). The 3D video clients are characterized by different median signal-to-noise ratios (E_S/N_0). LDPC codes are adopted, together with Adaptive Modulation and Coding, with a target frame error rate of 10^{-2}.

In all cases, "*Slice copy*" is enabled during decoding to conceal the missing slices of the corrupted bit-stream. At each $PLR/(E_S/N_0)$ and QP value, the quality is measured using the *FR SSIM* and the proposed methods for depth map and corresponding color images.

In order to show the degree of correlation between the proposed quality ratings and the true 3D video perception, subjective tests are performed using the double stimulus impairment scale (DSIS) method with the participation of 16 observers. 47" LG display with polarized glasses is utilized for these tests. Processed color and depth map sequences are rendered into left and right views using the DIBR method according to MPEG informative recommendations [36]. The dis-occluded regions (visual holes) are filled by background pixel extrapolation method, described in [37]. Standard test room under normal room lighting condition is selected for the tests. Subjects are screened for visual acuity (using the *Snellen* chart), color blindness (using the *Ishihara* test), and stereo vision (using the *Butterfly* Stereo test) prior to the tests. Two training sessions are conducted to familiarize with the 3D display and the test sequences used in the subjective tests. Subjective quality (i.e., overall 3D video quality) is measured for all the sequences under different *PLR*s. The sequences generated with $QP = 10$ and $QP = 30$ are selected for these tests. The opinion scores (*OS*) of individual subjects are averaged to obtain mean opinion score (*MOS*) for all the test cases. In addition to subjective quality ratings, the relationship between the proposed method and average PSNR quality of rendered left and right views is also demonstrated. This will also show the degree of correlation between the proposed quality ratings and the quality rating for rendered views using DIBR.

Figure 9.5 shows the quality measured with the proposed and *SSIM* quality metrics for color and depth map images of the *Ballet* sequence at $QP = 30$ (Figure 9.5a color image quality @ all PLRs and Fig. 9.5b depth map quality @ all PLRs). It can be clearly seen that both measures are closely matched for both color and depth map quality measurements. In order to minimize the offset between the proposed and the *FR* quality ratings, a relationship can be derived between the *FR* (SSIM$_D$ and SSIM$_C$) and the proposed methods (i.e., E-SSIM$_C$ and E-SSIM$_D$) based on experimental data using (9.12) and (9.13).

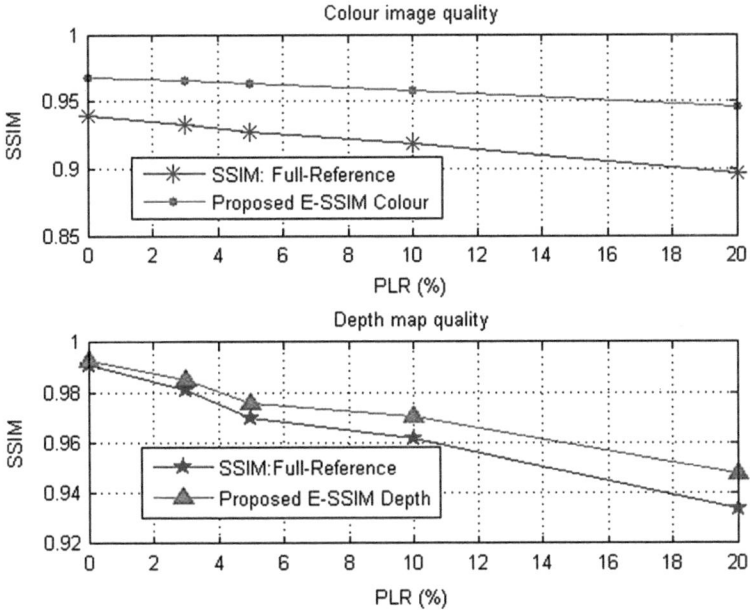

Fig. 9.5 Measured image quality of *Ballet* sequence at $QP = 30$ using *FR* and proposed *Near NR* method. (**a**) color image quality, (**b**) depth map quality

9.3.1   SSIM Based Near NR Quality Evaluation ("E-SSIM$_C$" and "E-SSIM$_D$" Methods)

Figures 9.6a–e show the scatter plots for the measured image quality (i.e., quality of both color image and depth map) using the *FR* and the proposed *Near NR* methods for the *Orbi, Interview, Ballet, Breakdance sequences*, and all the sequences in general respectively. Each point of this plot corresponds to the measured quality of an image frame using both *FR* and proposed method for a given compression level (QPs ranging from 1 to 50), *PLR* (0, 3, 5, 10, and 20 % PLRs), and E_S/N_0. According to these figures it is evident that there is a close relationship between the quality ratings of these methods regardless of the compression and PLR/(E_S/N_0) level being considered. In order to analyze the correlation between the reference and the proposed methods, the quality ratings are approximated by a second order polynomial [see Eqs. (9.14) and (9.15)] for all the sequences at different compression levels and packet loss rates. Fitting curves for the proposed and *FR* methods are shown in each of the scatter plots in Fig. 9.6 to show the quality of the fitting. This shows a higher degree of correlation between the proposed method and *FR* method for both color and depth map images.

$$\text{SSIM}_C \approx p_1 (\text{E-SSIM}_c)^2 + p_2 (\text{E-SSIM}_c) + p_3 \qquad (9.14)$$

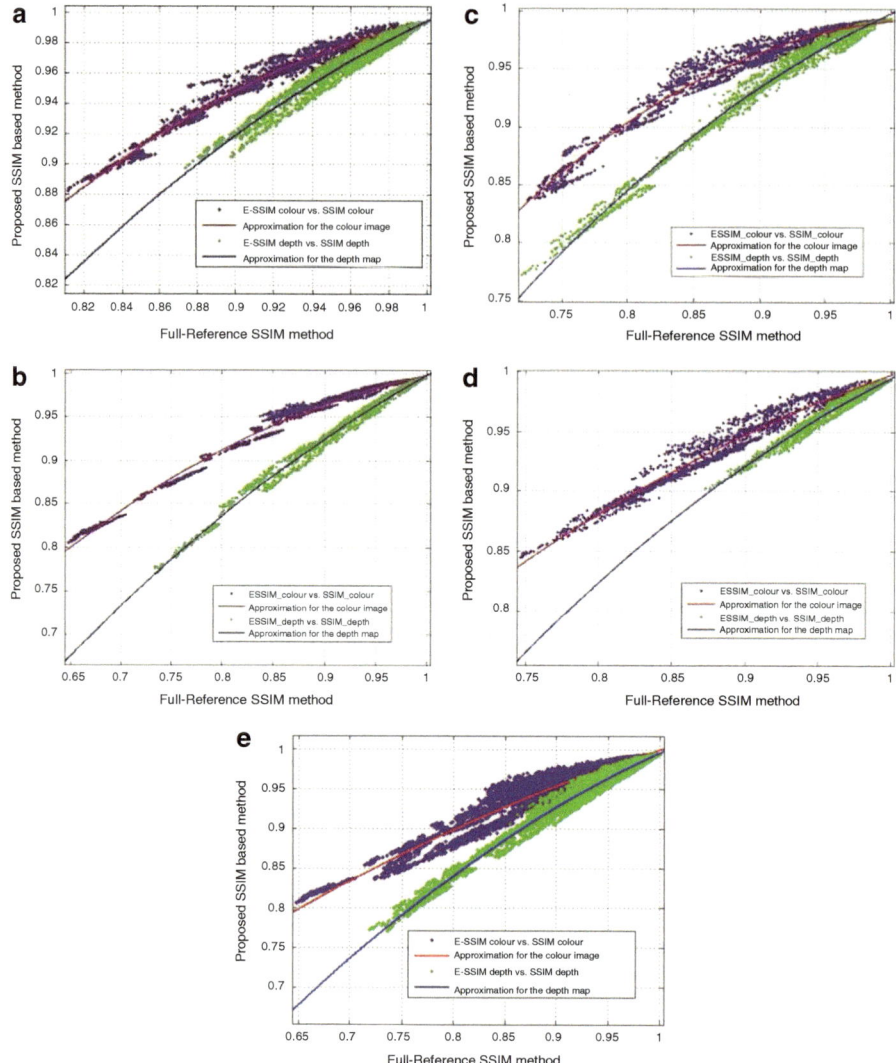

Fig. 9.6 Scatter plots of the proposed *Near NR* method versus *FR* method at different compression levels and PLR/(E_S/N_0) (**a**) *ballet* sequence, (**b**) *interview* sequence, (**c**) *orbi* sequence, (**d**) *breakdance* sequence, and (**e**) all the sequences in general

$$\text{SSIM}_D \approx p_1 \left(\text{E-SSIM}_D\right)^2 + p_2 \left(\text{E-SSIM}_D\right) + p_3 \qquad (9.15)$$

where p_1, p_2, *and* p_3 are constants and they have different values based on (9.14) and (9.15).

Figure 9.6e shows the scatter plot for the measured image quality (i.e., quality of both color image and depth map) using the *FR* and the proposed *Near NR* methods

Table 9.1 Performance indicators for different QP values

Sequence	Image sequence	SSE	R-square	RMSE	Spearman rank
Orbi	Color	0.12	0.97	0.01	0.98
	Depth	0.09	0.99	0.01	0.99
Interview	Color	0.06	0.99	0.00	0.99
	Depth	0.07	0.99	0.00	0.99
Ballet	Color	0.08	0.95	0.01	0.98
	Depth	0.05	0.95	0.00	0.97
Breakdance	Color	0.10	0.97	0.01	0.98
	Depth	0.03	0.97	0.00	0.98
All the sequences	Color	1.08	0.94	0.01	0.97
	Depth	0.37	0.98	0.01	0.98

for all the sequences in general. Each point of this plot corresponds to the measured quality of an image frame using both the *FR* and the proposed method for a given sequence, compression level, and $PLR/(E_S/N_0)$. Quadratic approximations using (9.14) and (9.15) are also plotted on the scatter plot. Readers should note that these results are achieved with a few bytes of overhead (to transmit *luminance* and *contrast* statistics of the original image sequence) compared to the *FR* method.

Individual performance indicators (i.e., SSE, Pearson correlation coefficient (as *R-square*), RMSE, and Spearman correlation coefficient) for the *Orbi*, *Interview*, *Ballet*, *Breakdance* sequences and for all the sequences in general are listed in Table 9.1. According to Table 9.1, it is clear that the proposed *Near NR* method (i.e., both "$E\text{-}SSIM_C$" and "$E\text{-}SSIM_D$") is highly correlated with the *FR* method at all QP levels, PLRs, and E_s/N_0 values considered. This suggests that we can use the proposed method for quality evaluation of the color image and of corresponding depth map in place of the *FR* method with a high accuracy regardless of the sequence type, compression level, and channel condition being used. This allows developers to use the proposed method to measure color plus depth based 3D video quality at the receiver "on the fly" with a very lower overhead for side-information. Based on the obtained individual quality ratings for color and depth map, receiver-side can provide feedback to the sender-side about current status of receiving sequences for system parameter optimization and perhaps receiver can take decisions such as whether to display 2D content or 3D content. For instance if the measured depth map quality is significantly low, the quality of the rendered views will be low. Therefore, the receiver can decide to shift back to 2D video mode instead of 3D video.

In order to better understand the degree of correlation between true 3D perception (i.e., subjective quality ratings) and results obtained for color and depth map image quality evaluation with the proposed method, the objective results are mapped with the subjective MOS results. The mapping between the proposed *Near NR* metric for color images (i.e., "$E\text{-}SSIM_C$" ratings) and subjective quality ratings for "overall 3D image quality" is shown in Fig. 9.7. This scatter plot shows the average subjective quality (*MOS*) and proposed objective quality ratings for the color image ("$E\text{-}SSIM_C$") under different $PLR/(E_S/N_0)$ for all four sequences

Fig. 9.7 Scatter plot of subjective MOS for overall 3D image quality versus color image quality obtained with the proposed "$E\text{-}SSIM_C$" method. Each sample point represents one test sequence at a given $PLR/(E_S/N_0)$

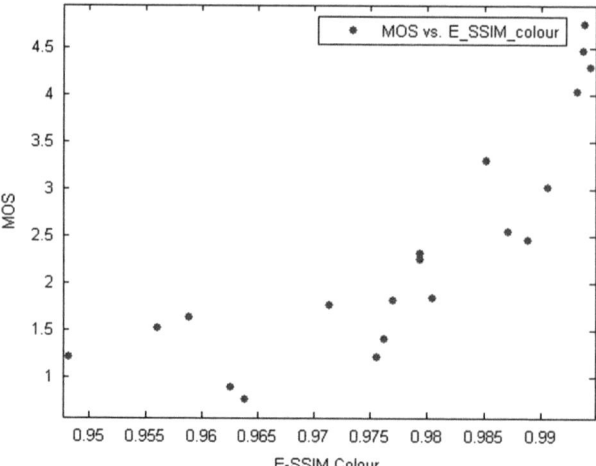

Fig. 9.8 Scatter plot of subjective MOS for overall 3D image quality versus depth map quality obtained with the proposed "$E\text{-}SSIM_D$" method. Each sample point represents one test sequence at a given $PLR/(E_S/N_0)$

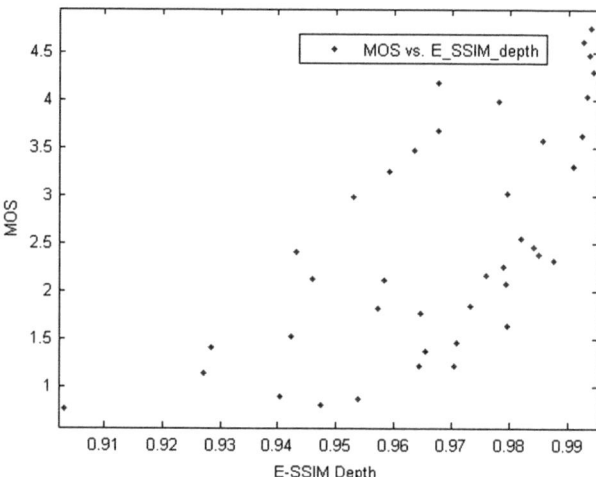

(compressed with $QP = 10$ and $QP = 30$). Figure 9.8 shows the mapping between proposed method for depth map quality evaluation (i.e., "$E\text{-}SSIM_D$" ratings) and subjective MOS results. Sequences encoded using both $QP = 10$ and $QP = 30$ are used for this mapping. Since average quality ratings are shown for all test points, only a few number of test points are visible in the scatter plot. This is mainly due to the fact that subjective quality ratings are obtained for the whole image sequence (and averaged across all the subjects) and not for each frame. These scatter plots show the degree of correlation between proposed objective "*Near NR*" measures and subjective MOS. However, it is difficult to derive a straight forward relationship between these objective measures of color and depth map with the true 3D video perception due to their usage in viewing and rendering. For instance, depth map is not directly viewed by the users and only used for calculating pixel shift for left and right views. However, quality measures for color image show some degree of direct relationship with the true user perception (i.e., subjective ratings for overall 3D video quality). This is mainly due to the fact that color image artifacts are directly visible in rendered left and right views. However, artifacts in the depth map also add some degree of distortion to the rendered left and right views. More details on the relationship between *FR* objective metrics and subjective scores for 3D video sequences can also be found in [15].

The rendered quality of left and right views can also be used as an overall quality metric for 3D video quality. Figures 9.9 and 9.10 show the degree of correlation

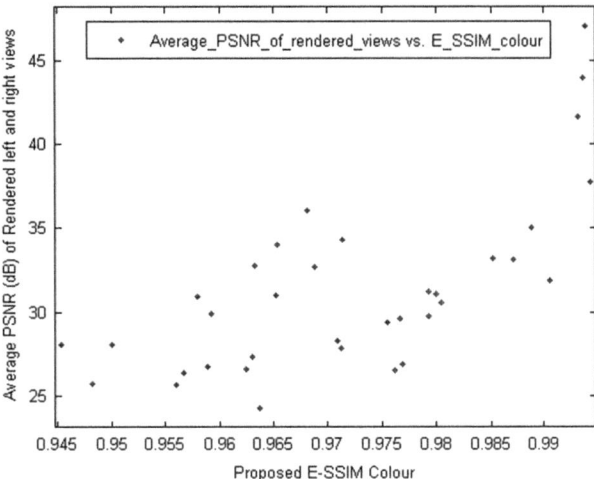

Fig. 9.9 Scatter plot of average PSNR (dB) of the rendered left and right image versus color image quality obtained with the proposed "*E-SSIM$_C$*" method. Each sample point represents one test sequence at a given *PLR*/(E_S/N_0)

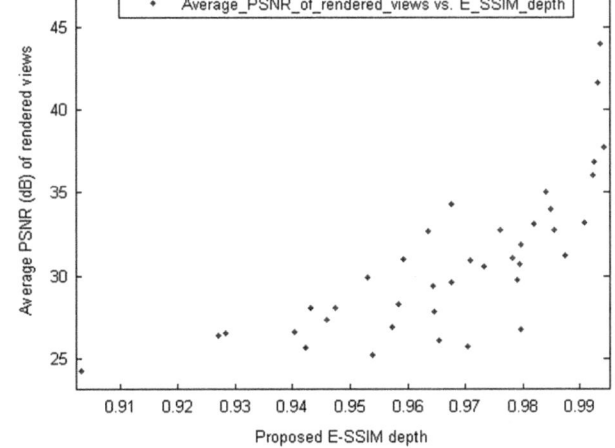

Fig. 9.10 Scatter plot of average PSNR (dB) of the rendered left and right image versus depth map quality obtained with the proposed "*E-SSIM$_D$*" method. Each sample point represents one test sequence at a given *PLR*/(E_S/N_0)

between the proposed *Near NR* quality metric versus average PSNR (dB) of rendered left and right view quality for color and depth map sequences respectively. These figures consider sequences encoded at both $QP = 10$ and $QP = 30$. These scatter plots also demonstrate a higher degree of correlation between the proposed quality evaluation method and the rendered quality of left and right views with DIBR. These findings suggest that the proposed method can be used to measure the quality of individual color and depth map views or approximation of true 3D perception (MOS) on-the-fly with a minimum overhead for side-information.

Conclusion

This chapter proposed a *NR* quality metric based on edge detection for color plus depth 3D video compression and transmission. Since the edges of the color image and the corresponding depth maps represent different depth levels and the basic structure of the color and depth map objects, edge information extracted at the receiver side is used as a measure to quantify the structural degradation of these sequences. Since only statistics related to *luminance* and *contrast* information of the original image sequences are required for image quality evaluation as side-information, the proposed method is close to a complete *NR* quality metric, hence referred to as *Near NR* in this chapter. Results show a good approximation of the associated *FR* quality metric for all considered *PLR*s, compression levels, and wireless impairments. However, due to the abstract level of information used to measure the structural degradation with the proposed method, results may differ from the *FR* method. In order to obtain matched results, the proposed method can be calibrated against the *FR SSIM* metric based on experimental results. In this case the performance is almost indistinguishable from the *FR* method, with an R-square value of 0.94 and 0.98 for color and depth map sequences respectively.

This suggests that due to the practical problems associated with the usage of *FR* methods for online system optimization, *NR* quality metrics as described in this chapter are an acceptable compromise for the design of QoE-aware 3D multimedia systems.

References

1. Le-Callet P, Moeller S, Perkis A (2012) Qualinet white paper on definitions of quality of experience, Version 1.1. European network on quality of experience in multimedia systems and services, COST Action IC 1003, June, 2012
2. Martini MG, Mazzotti M, Lamy-Bergot C, Huusko J, Amon P (2007) Content adaptive network aware joint optimization of wireless video transmission. IEEE Commun Mag 45 (1):84–90
3. Meesters LMJ, IJsselsteijn WA, Seuntiens PJH (2004) A survey of perceptual evaluations and requirements of three-dimensional TV. IEEE Trans Circuits Syst Video Technol 14(3):381–391

4. Wang K, Barkowsky M, Brunnstrom K, Sjostrom M, Cousseau R, Le-Callet P (2012) Perceived {3D TV} transmission quality assessment: multi-laboratory results using absolute category rating on quality of experience scale. IEEE Trans Broadcast 58(4):544–557
5. Cutting JE, Vishton PM (1995) Perceiving layout and knowing distances: the integration, relative potency, and contextual use of different information about depth. In: Epstein W, Rogers S (eds) Perception of space and motion. Academic, San Diego, pp 69–117
6. Lambooij M, Fortuin M, Heynderickx I, IJsselsteijn W (2009) Visual discomfort and visual fatigue of stereoscopic displays: a review. J Imaging Sci Technol 53(3):30201-1–30201-14
7. Tam W, Stelmach L, Corriveau P (1998) Psychovisual aspects of viewing stereoscopic video sequences. Proc SPIE 3295:226–235
8. IJsselsteijn W, De Ridder H, Vliegen J (2000) Subjective evaluation of stereoscopic images: effects of camera parameters and display duration. IEEE Trans Circuits Syst Video Technol 10 (2):225–233
9. International Telecommunication Union/ITU Radio communication Sector (2000) Subjective Assessment of Stereoscopic Television Pictures. ITU-R BT.1438, Jan 2000
10. International Telecommunication Union/ITU Radio communication Sector (2002) Methodology for the subjective assessment of the quality of television pictures. ITU-R BT.500-11, Jan 2002
11. Seuntiens P, Meesters L, Ijsselsteijn W (2008) Perceived quality of compressed stereoscopic images: effects of symmetric and asymmetric JPEG coding and camera separation. ACM Trans Appl Percept (TAP) 3(2):95–109
12. Seuntians P, Meesters L, IJsselsteijn W (2003) Perceptual evaluation of JPEG coded stereoscopic images. Proc SPIE 5006:215–226
13. Starck J, Kilner J, Hilton A (2008) Objective quality assessment in free-viewpoint video production. In: 3DTV conference, Turkey
14. Benoit A, Le Callet P, Campisi P, Cousseau R (2008) Quality assessment of stereoscopic images. EURASIP J Image Video Proc 2008:1–13
15. Hewage CTER, Worrall ST, Dogan S, Villette S, Kondoz AM (2009) Quality evaluation of color plus depth map-based stereoscopic video. IEEE J Sel Top Sign Proces 3(2):304–318
16. Hewage CTER, Worrall ST, Dogan S, Kondoz AM (2008) Prediction of stereoscopic video quality using objective quality models of 2-D video. IET Electron Lett 44(16):963–965
17. Yasakethu SLP, Hewage CTER, Fernando WAC, Worrall ST, Kondoz AM (2008) Quality analysis for 3D video using 2D video quality models. IEEE Trans Consumer Electron 54 (4):1969–1976
18. Babu RV, Bopardikar AS, Perkis A, Hillestad OI (2004) No-reference metrics for video streaming applications. In: Proc Int Packet Video Workshop
19. Wang Z, Sheikh HR, Bovik AC (2002) No-reference perceptual quality assessment of JPEG compressed images. In: Proc IEEE Int. Conf. on Image Processing. pp 477–480
20. Wang Z, Simoncelli E (2005) Reduced-reference image quality assessment using a wavelet-domain natural image statistic model. In: 17th SPIE annual symposium on electronic imaging, San Jose, Jan 2005
21. Wolf S, Pinson MH (2005) Low bandwidth reduced reference video quality monitoring system. In: First inter. workshop on video processing and quality metrics for cons electronics, Arizona, Jan 2005
22. Akhter R, Sazzad ZMP, Horita Y, Meek D (2010) No reference stereoscopic image quality assessment. In: Proc of SPIE: stereoscopic displays and applications XXI 7524, California, Jan 2010
23. Merkle P, Smolic A, Muller K and Wiegand T (2007) Multi-View Video Plus Depth Representation and Coding. In: IEEE Int Conf Image Processing, San Antonio, pp 201–204
24. Fehn C (2003) A 3D-TV approach using depth-image-based rendering (DIBR). In: Proceedings of Visualization, Imaging, and Image Processing (VIIP'03), pp 482–487
25. Lee C, Cho S, Choe J, Jeong T, Ahn W, Lee E (2006) Objective video quality assessment. Opt Eng 45(1):017004–017004

26. Chen GH, Yang CL, Xie SL (2006) Gradient-based structural similarity for image quality assessment. In: Int Conf Image Processing, IEEE. pp 2929–2932
27. Hewage CTER, Martini MG (2010) Reduced-reference quality metric for 3D depth map transmission. In: 3DTV conference 2010, Finland
28. Hewage CTER, Martini MG (2010) Reduced reference quality metric for compressed depth maps associated with colour plus depth 3D video. In: IEEE Int Conf Image Processing (ICIP 2010), September 26–29, Hong Kong
29. Ekmekcioglu E, Mrak M, Worrall ST, Kondoz AM (2009) Edge adaptive up-sampling of depth map videos for enhanced free-viewpoint video quality. IET Electron Lett 45(7):353–354
30. Wenger S (1999) Error patterns for Internet video experiments. ITU-T SG16 Document Q15-I-16-R1
31. Wang Z, Bovik AC, Sheikh HR, Simoncelli EP (2004) Image quality assessment: from error measurement to structural similarity. IEEE Trans Image Process 13(4):600–613
32. Chen GH, Yang CL, Po LM, Xie SL (2006) Edge-based structural similarity for image quality assessment. In: Int conf acoustics, speech and signal processing. pp 933–936
33. Marr D, Hildreth E (1980) Theory of edge detection. In: Proc of the Royal Society of London. Series B
34. Woods J (2006) Multidimensional signal, image and video processing and coding. Academic, Elsevier
35. Kazakova N, Margala M, Durdle NG (2004) Sobel edge detection processor for a real-time volume rendering system. In: Proc Int Symp Circuits Syst. pp 913–916
36. ISO/IEC JTC 1/SC 29/WG 11 (2006) Committee Draft of ISO/IEC 23002-3 Auxiliary Video Data Representations, Doc. N8038
37. Gangwal O, Berretty R (2009) Depth map post-processing for 3D-TV. Consumer electron (ICCE '09), pp 1–2

Chapter 10
Visual Discomfort in 3DTV: Definitions, Causes, Measurement, and Modeling

Jing Li, Marcus Barkowsky, and Patrick Le Callet

Abstract This chapter discusses the phenomenon of visual discomfort in stereo-scopic and multi-view 3D video reproduction. Distinctive definitions are provided for visual discomfort and visual fatigue. The sources of visual discomfort are evaluated in a qualitative and quantitative manner. Various technological influence factors, such as camera shootings, displays, and viewing conditions are considered, providing numerical limits for technical parameters when appropriate and available. Visual discomfort is strongly related to the displayed content and its properties, notably the spatiotemporal disparity distribution. Characterizing the influence of content properties requires well-controlled subjective experiments and a rigorous statistical analysis. An example of such a study for measuring the influence of 3D motion is presented in detail and important conclusions are drawn.

10.1 Definitions of Visual Discomfort and Visual Fatigue in 3D

Visual discomfort and visual fatigue are two distinct concepts though they are often confused and interchangeably used in some papers. Visual fatigue and discomfort are notions that encompass medical, psychological, and subjectively perceived aspects. Accordingly, related studies are found in between associated research fields. For this reason, terminologies and definitions may vary from one study to the other and are sparsely explained.

Before introducing the definitions of visual fatigue and discomfort, some terms should be defined first.

- Symptom: A symptom is a subjective sensation reported by the patient, as an evidence of his perceived physical or mental condition. Symptom is subjectively experienced, it cannot be measured directly [1].
- Clinical sign: A clinical sign is observed or measured by the medical examiner, thus is an objective evidence of a patient's condition [2].

J. Li (✉) • M. Barkowsky • P. Le Callet
LUNAM Université, Université de Nantes, IRCCyN UMR CNRS 6597, Polytech Nantes, rue Christian Pauc BP, 50609 44306 Nantes, Cedex 3, France
e-mail: jing.wang@univ-nantes.fr

© Springer Science+Business Media New York 2015
A. Kondoz, T. Dagiuklas (eds.), *Novel 3D Media Technologies*,
DOI 10.1007/978-1-4939-2026-6_10

– Syndrome: A syndrome is a set of (subjective) symptoms and (objective) signs that occur together, which is characteristic of a physical or mental condition [3].

10.1.1 Visual Fatigue

Depending on the context, fatigue is either considered as a symptom of a medical condition, or a medical condition itself. In our context, the latter terminology applies.

Visual fatigue is caused by the repetition of excessive visual efforts, which can be accumulated, and disappears after an appropriate period of rest. Visual fatigue can be assessed by the presence of [4].

– Zero, one or more symptoms reported by the patient, which may include the sensation of fatigue reported by the patient;
– Zero, one or more clinical signs observed by the medical examiner or measured through experimental protocols.

Their nature, intensity, and temporal properties (time of appearance, duration, raise, and fall time) may be used to assess the severity of visual fatigue.

10.1.2 Visual Discomfort

Visual discomfort is a physical and/or a psychological state assessed by the patients themselves, as a presently perceived degree of annoyance. As such, it may be related to experienced symptoms, perceived difficulties when performing a visual task or any negative sensation associated with this task. Visual discomfort appears and disappears with any of these negative associations, and is supposed to have a short raise and fall time compared to visual fatigue. In other words, visual discomfort disappears rapidly when the visual task is interrupted, either by asking the observer to close his eyes or by terminating the visual stimulus. Visual discomfort can be measured by asking the viewer to report its level [4]. In this chapter, we focus on the visual discomfort issues.

10.2 Main Causes of Visual Discomfort

10.2.1 Vergence–Accommodation Conflict

Vergence–Accommodation conflict is a well-known factor that would induce visual discomfort [5, 6]. When viewing an object by means of a 3D screen, the eyes will converge to the virtual object which is in front of or behind the screen plane. However, the accommodation has to be performed at the screen itself, i.e. with respect to the distance of the observer in front of the screen, which is unnatural and will not happen in our daily life. The larger this discrepancy between the vergence and accommodation gets, the higher the possibility that observers will perceive visual discomfort.

To define the threshold of this discrepancy in which conditions viewers may not experience visual discomfort, i.e., the comfortable viewing zone, numerous studies have been conducted. Yano et al. [7] proposed that the depth of field (DOF), which refers to the range of distances in image space within which an image appears in sharp focus, can be used to define the comfortable viewing zone in terms of diopters (D). A value of ±0.2 D is suggested [8, 9]. Another definition on comfortable viewing zone is based on the results of empirical measurements, in which ±1 arc degree of visual angle is used [10, 11]. If considering the screen disparity, the comfortable viewing zone can be defined by a percentage of the horizontal screen size. For 3D television, values of ±3 % are suggested [12]. For cinema, values of 1 % for crossed and 2 % for uncrossed disparities are suggested [13].

Generally, these definitions generate similar 3D spaces of comfortable viewing [14].

10.2.2 Disparity Distribution

In addition to the Vergence–Accommodation conflict, some studies also showed that the disparity distribution might introduce visual discomfort as well:

1. Excessive uncrossed disparity (behind screen) will induce less visual discomfort compared to the crossed disparity (in front of the screen) when the angular disparity magnitude is the same [15].
2. When most parts of an image are positioned behind the screen (or the averaged disparity is uncrossed), there will be less visual discomfort compared with the condition that they are distributed in front of the screen [16].
3. If the image is split into top and bottom parts, the stereoscopic image will be more comfortable to watch when the top part of the image is distributed behind the screen and the bottom of the image is in front of the screen [17].
4. In the condition of the same averaged value of disparity distribution, higher dispersion of the disparity would lead to more visual discomfort due to the Vergence–Accommodation conflict [16].

10.2.3 Binocular Distortions

Binocular distortions or binocular image asymmetries seriously reduce visual comfort if present to a sufficient extent [18]. Asymmetries can be classified into optics related errors, filters related errors, and display related errors. Optics errors are mainly geometry differences between the left and right images, e.g., size inconsistency, vertical shift, rotation error, magnification, or reduced resolution. These errors usually occur when shooting or displaying stereoscopic images/videos. Filters related errors are mainly photometry differences between the two views, e.g., color, sharpness, contrast. The main error induced by display systems is crosstalk. Crosstalk produces double contours and is a potential cause of discomfort [19]. A study showed that vertical disparity, crosstalk, and blur are most dominant factors when compared with other binocular factors in visual comfort [18].

10.2.3.1 Optics Related Errors

The accuracy of the alignment between the two cameras during a stereoscopic shooting determines the perceptual visual comfort to a large extent. Generally, the camera misalignments are divided into [20]:

- Horizontal misalignments;
- Vertical misalignments;
- Torsional misalignments;
- Size and keystone disparity fields.

There are three types of camera misalignment: The pitch and yaw axis horizontally and vertically through the picture plane, while the roll axis coinciding with the optic axis through the center of the lens. Vertical inconsistency of images caused by inconsistency of optic axes and errors in the rotational alignment between the two cameras are known to cause fatigue of eyes. Studies in [20] indicated that the relationship between visual discomfort, vertical disparity (in unit of arcmin), and torsional disparity (degree) are that, in the condition of watching 3D stimuli for 2 s, vertical disparity about 60 arcmin will induce severe visual discomfort (the five-level rating scale is: 0—no discomfort, 1— mild, 2—moderate, 3—strong, 4—severe discomfort). For torsional disparity, 50° will induce strong discomfort.

In the converged camera configuration there is often a distortion called keystone distortion. It is the phenomenon that in one of the views, the image of the grid appears larger at one side than the other as shown in Fig. 10.1 [21]. Studies already showed that this distortion will induce visual discomfort or even visual fatigue [22, 23].

The tolerance of these optics related errors on visual discomfort has been investigated in [24] and the results are shown in Table 10.1.

10.2.3.2 Filters Related Errors

In [25], the Just Noticeable Difference (JND) thresholds of the visual comfort in 3DTV were investigated psychovisually for seven types of between-eye image differences, including luminance, gamma, contrast, color temperature, chroma, hue, and random tone differences. The experimental results showed that: (1) the

Fig. 10.1 Keystone
distortion

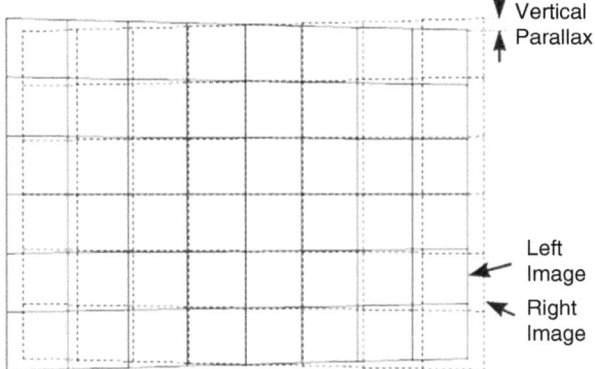

Vertical
Parallax

Left
Image

Right
Image

Table 10.1 Studies on the detection and tolerance limits of the geometric discrepancy on left and right views [24]

Geometric discrepancy	Detection limit	Tolerance limit	Remark
Size	1.2 %	2.9 %	Taking the size of one image as 100 %
Vertical displacement	0.7 %	1.5 %	Taking the image height as 100 %
Rotation	0.5°	1.1°	Angle of rotation about the image center

visual comfort threshold values are higher when increasing the luminance and color temperature differences between two views, (2) decreasing the binocular differences on contrast or hue to zero will result in low threshold values on visual discomfort, which indicates this type of differences easily inducing binocular rivalry, and (3) luminance adaptation and chromaticity adaptation play an important role on the variations of visual comfort thresholds.

10.2.3.3   Crosstalk

Crosstalk, or image ghosting, in stereoscopic displays refers to imperfect image separations where the left-eye view leaks to the right-eye view and vice versa. Crosstalk can be classified into system crosstalk and viewer crosstalk. System crosstalk is only induced by the display technology, and it is independent of the quality of stereoscopic image pairs while the viewer crosstalk is dependent on the video contents, for example displaying high contrast edges with non-zero disparity.
    Studies showed that visibility of crosstalk increases with increasing contrast and increasing binocular disparity (depth) of the stereoscopic image [26]. Even a small amount of crosstalk would induce visual discomfort or visual fatigue [19]. Studies on the thresholds of crosstalk level for the acceptance of the viewing experience and visual discomfort have been conducted. In [27], they found that "crosstalk between

2 and 6 % significantly affected image quality and visual comfort." In [18], they found "crosstalk level of about 5 % is sufficient to induce visual discomfort in half of the population." In [28], they reported that the crosstalk tolerance limit is 5–10 %, and visual detection limit is 1–2 %. In [29], it is shown with natural still images that the S-3D display technology with the lowest luminance and contrast level tolerates the highest level of crosstalk, while still maintaining an acceptable image-quality level.

10.2.4 Motion

Motion in 3DTV can be classified into planar motion (or lateral motion) and in-depth motion. Planar motion means that the object only moves in a certain depth plane perpendicular to the observer, and the disparity does not change temporally. In-depth motion, which is also called motion in depth or z-motion, is defined as object movement toward or away from an observer [30]. For planar motion, both eyes make the same conjunctive eye movements, called version [31]. For in-depth motion, the eyes make opposite, disjunctive eye movement, called vergence [32]. The eye movements for the planar motion and in-depth motion are shown in Fig. 10.2. The speed of the planar motion and in-depth motion can be expressed by the change of distance per second or the change of the visual angle (version or vergence) per second.

Fast motion can induce visual discomfort even if the object is within the comfortable viewing zone [9, 11]. Studies showed a consistent conclusion on the influence of motion velocity on visual discomfort, i.e., visual discomfort increases with the in-depth motion velocity [9, 11, 12, 31], and for planar motion video sequences, visual discomfort increases with the planar motion velocity [12, 33].

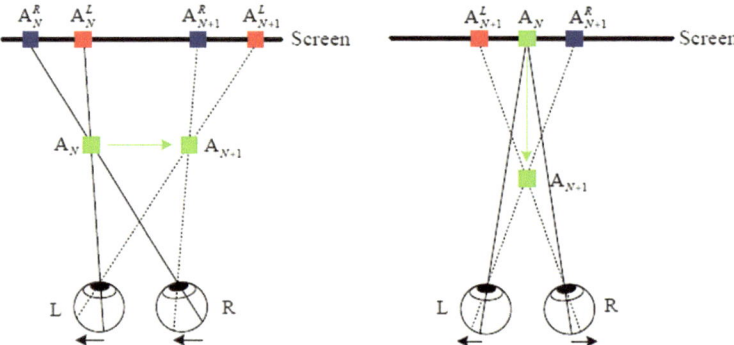

Fig. 10.2 The *left figure* shows the eye movement of the planar motion object. Planar motion velocity is the amount of the change of the version per second. The *right figure* shows the eye movement of the in-depth motion object. In-depth motion velocity is the amount of the change of the vergence per second. A_N represents the perceived virtual object at frame N, and A_N^L and A_N^R represent the *left* and *right* view images at frame N

10.3 Prediction of Visual Discomfort Induced by Motion

It is well accepted that large disparity and large amount of motion are two main causes of visual discomfort. To quantify this influence, it is necessary to design a subjective experiment on visual discomfort induced by disparity and motion. In this section, a well-controlled subjective experiment on 3D visual discomfort is introduced as an example of how to design and conduct a reproducible and reliable experiment in the area of 3D.

Three objectives are aimed at in the experiment. The first one is the comparative analysis on the influence of different types of motion on visual discomfort, namely static stereoscopic image, planar motion, and in-depth motion. The second one is the investigation on the influence factors for each motion type, for example, the disparity offset, the disparity amplitude, and velocity. The third one is to propose an objective model for visual discomfort. In addition, the influence from viewers' 3D experience, i.e., the differences between experts and naive viewers is introduced.

10.3.1 State of the Art

As we already introduced in Sect. 10.2, there are many possible factors that would induce visual discomfort. Most of the causes have been well studied for the case of still stereoscopic images. For stereoscopic 3D videos, as the only difference between stereoscopic image and video is the motion, the influence of motion on visual discomfort has been widely investigated recently.

In-depth motion is one of the significant factors that may cause visual discomfort. Studies already showed that visual discomfort increases with the in-depth motion velocity [9, 11, 12, 31, 33]. However, the influence from disparity amplitude (disparity range) and the disparity type (crossed or uncrossed) of in-depth motion on visual discomfort are still under study. In [11], the results showed that disparity amplitude of the moving object is not a main factor. However, in their recent study [12] it is shown that visual discomfort increases with the disparity amplitude. Furthermore, the results also showed that the in-depth motion with crossed disparity would induce significantly more visual discomfort than the uncrossed and mixed conditions. In [33], as they only analyzed the in-depth motion in the disparity range of $\pm 1°$ with different velocities, there is no conclusion about the influence of crossed or uncrossed disparity amplitude on visual discomfort.

The influence of the planar motion on visual discomfort was studied as well [12, 33–35]. These studies showed high consistency on the conclusion that visual discomfort increases with the motion velocity. However, the influence of the disparity on visual discomfort led to different conclusions in these studies. In [33], the results indicated that the disparity type, i.e., crossed and uncrossed disparity, did not affect the visual discomfort thresholds. However, in [1], the results showed that the crossed disparity will generate more visual discomfort than the uncrossed disparity. A possible explanation for this inconsistency might be the position of the background. In [33], the background was positioned at the screen plane. In [12], the position of the background was not depicted but in their previous study [11], the background was positioned at a fixed place with the disparity of— 2.6°. The impact of the position of background on visual discomfort may therefore require further study.

Most of the studies mentioned above investigated the influence of the in-depth motion and planar motion on visual discomfort individually. For quantifying the influence of static situations, planar and in-depth motion, it would be important to directly compare their impact on visual discomfort. Thus, this section is focusing on the influence of motion on visual discomfort of 3DTV, including the comparative analysis on the influence of different motion types on visual discomfort, the influence of disparity and velocity within a certain motion type, and the proposal of an objective visual discomfort model based on the results. In addition, the influence of viewers' experience (experts and naive observers) on 3DTV is analyzed.

10.3.2 Experiment

Based on the two main objectives, i.e., analysis on the influence from motion and the study on the influence from human factors, two experiments were designed as shown in Table 10.2.

Table 10.2 Summary of the two experiments

Item	Exp 1	Exp 2 Exp2-a	Exp2-b
Study target	Influence from motion	Human factors	
Number of stimuli	36 (static + planar + in-depth)	15 (planar)	
Method	ASD	FPC	
Display technology	Shutter glasses		
Resolution	Full HD, 1920 × 1080		
Viewing distance	3H, 90 cm		
Number of observer	42	10	45
Observer type	Naive observer	Experts	Naive observer
Gender	21 males + 21 females	8 males + 2 females	21 males + 24 females
Age (mean)	19–48 (26.8)	24–43 (27)	18–44 (24)
Trials/obs	180 pairs	210 pairs	105 pairs
Votes/stimulus pair	42	15	45

10.3.2.1   Definitions

To analyze the influence of crossed and uncrossed disparity, and the disparity
magnitude of the moving object on visual discomfort, in this study, we use *disparity
amplitude* and *disparity offset* to define the motion in stereoscopic videos. The
disparity amplitude d_a between the nearest point A and farthest point B can be
expressed by Eq. (1) which represents the range of the disparity of the moving
object. ϕ_A represents the disparity of point A, and ϕ_B represents the disparity of
point B. The disparity offset d_o between the two points A and B can be expressed by
Eq. (2) which represents the center of the angular disparity between the two points.

The static and planar motion stimuli can be characterized by the disparity offset
and the planar velocity, where the disparity amplitude equals zero. The in-depth
motion stimuli can be defined by the disparity amplitude, the disparity offset, and
the in-depth velocity.

$$d_a = |\phi_A - \phi_B| \tag{10.1}$$

$$d_o = \frac{1}{2}(\phi_A + \phi_B) \tag{10.2}$$

10.3.2.2   Experimental Design

To avoid the complexity of the influence factors contained in 3D natural video
sequences, synthetic stimuli were used in this study allowing for precise control on
the possible influence factors, including motion type, velocity, disparity offset, and
disparity amplitude. In Experiment 1, 36 synthetic video stimuli were used, including 15 planar motion stimuli, 5 static stimuli, and 16 in-depth motion stimuli.

For the planar motion stimuli, we selected five angular disparity offset levels
(0, ±0.65, and ±1.3°) and three velocity levels (slow, medium, and fast with
velocity of 6, 15, and 24°/s). A background is designed to be placed at a fixed
position (−1.4°) which is consistent with a typical natural video content where the
background is almost fixed and placed behind the screen. Figure 10.3 shows the
disparities used in the planar motion stimuli and their relationship with the comfortable viewing zone.

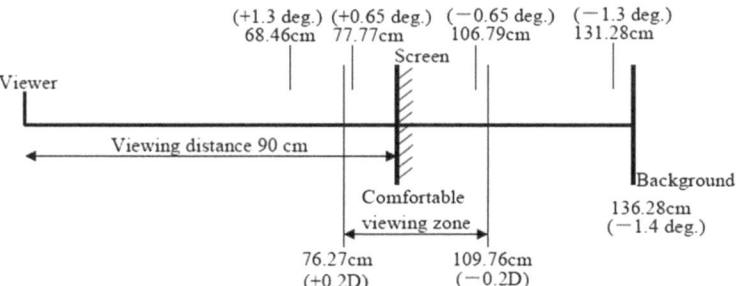

Fig. 10.3 The relationship of the foreground and the background position and the comfortable
viewing zone in planar motion stimuli

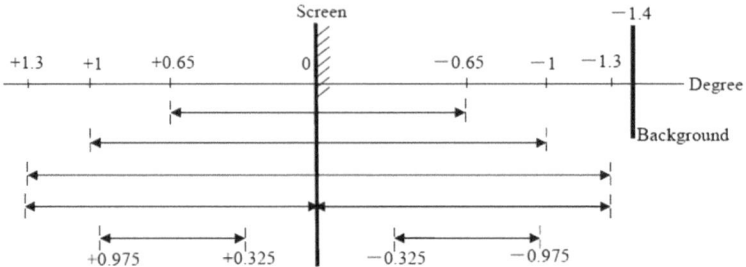

Fig. 10.4 The disparity amplitude and offset design for in-depth motion stimuli. The *arrows* represent the depth interval in which the object moves

For the static condition, five disparity offset levels were selected which were the same as the planar motion design. The foreground object stays fixed in the center of the screen at a certain disparity plane.

For the in-depth motion condition, four disparity amplitude levels (0.65, 1.3, 2, and 2.6°), three disparity offset levels (−0.65, 0, 0.65°), and three velocity levels (1, 2, and 3°/s, binocular angular degree) were selected. The reason for choosing binocular angular disparity speed was that the object's velocity appears visually constant, which is not the case for a constant value in the unit of cm/s. The direction of the movement is inverted at the far or at the near end of the movement so the object in the experiment moved forth and back in an endless loop. The three velocity levels 1, 2, and 3°/s represent slow, medium, and fast, respectively. Figure 10.4 shows the design of the disparity amplitude and disparity offset for the in-depth motion.

In experiment 2, only the 15 planar motion stimuli were used. Two types of viewers participated in the test, one being expert viewers who work in 3D perception, coding, quality assessment, and subjective experiments and are thus familiar with 3D depth perception, visual discomfort, etc. The others were naive viewers, who do not have experience in this domain.

10.3.2.3 Apparatus

It is important to document the apparatus used in the experiment for the reproduction of the same research topic or the comparison of the similar studies by others.

In this experiment, the stereoscopic sequences were displayed on a Dell Alienware AW2310 23-inch 3D LCD screen (1920 × 1080 full HD resolution, 120Hz), which featured 0.265-mm dot pitch. The display was adjusted for a peak luminance of 50 cd/m² when viewed with the active shutter glasses. The graphics card of the displaying PC was an NVIDIA Quadro FX 3800. Stimuli were viewed binocularly through the NVIDIA active shutter glasses (NVIDIA 3D vision kit) at a distance of about 90 cm, which was approximately three times the picture height. The peripheral environment luminance was adjusted to about 44 cd/m². When seen through the eye-glasses, this value corresponded to about 7.5 cm/m^2 and thus to 15 % of the screen's peak brightness as specified by ITU-R BT.500 [36].

10.3.2.4   Stimuli

The stereoscopic sequences consisted of a left-view and a right-view image which were generated by the MATLAB psychtoolbox [37, 38]. Each image contained a foreground object and a static background. A black Maltese cross which was frequently used in such kind of psychometric experiments [39, 40] was used as the foreground object with a resolution of 440×440 pixels corresponding to a visual angle of 7.6°. As it contained both high and low spatial frequency components, it was supposed to limit the influence of one particular spatial frequency in the experiment [41].

The background was generated by adding salt and pepper noise to a black image of Full HD resolution, and then filtered by a circular averaging filter with radius of 5. The reason for using this kind of image as the common background of all stimuli was that it could preclude all of the monocular cues on stereopsis.

For the planar motion stimuli, the trajectory of the moving object is a circle with center point at the center of the screen, and radius of 300 pixels, approximately 10° of visual angle. The motion direction of the object was anti-clockwise. An example of the stimuli is shown in Fig. 10.5a, in which the foreground object is placed in front of the screen with an angular disparity of 1.3°. For the static stimuli, the Maltese cross was positioned at the center of the screen. For the in-depth motion stimuli, the Maltese cross was positioned in the center of the screen and moved back and forth to the viewers. An example is shown in Fig. 10.5b, in which the foreground object is moving in the depth plane with disparity amplitude of 2.6° and offset of 0°.

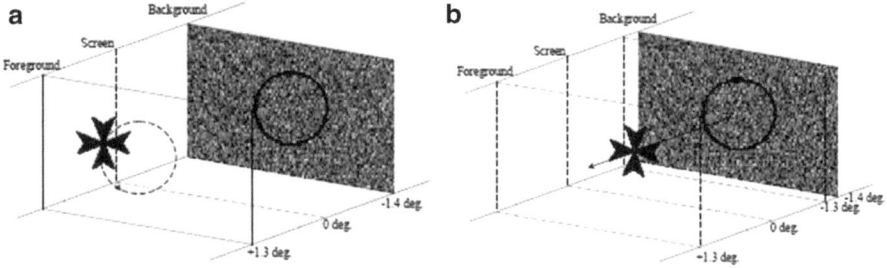

Fig. 10.5 (a) An example of stimulus with planar motion in the experiment. The foreground object is moving at the depth plane with a disparity of 1.3°. The background is placed at a fixed depth plane of −1.4°. The motion direction of the Maltese cross is anti-clockwise. (b) An example of stimulus with in-depth motion in the experiment. The disparity amplitude of the Maltese cross is 2.6°, and offset is 0°. The foreground object is moving in depth between disparities +1.3 and −1.3° back and forth

The 15 planar motion stimuli, 5 static stimuli, and 16 in-depth motion stimuli used in the subjective experiment are listed in Table 10.3 with their stimulus serial number, disparity offset d_o, disparity amplitude d_a, planar motion velocity v_p, and in-depth motion velocity v_d.

Table 10.3 All stimuli used in the experiment. d_o is disparity offset, d_a is disparity amplitude, v_p is planar motion velocity, and v_d is in-depth motion velocity. The last two columns are BT score and confidence interval of the BT score (CI)

Type	Number	d_o (deg.)	d_a (deg.)	v_p (deg./s)	v_d (deg./s)	BT score	CI
Planar motion	1	-1.3	0	6	0	0	0.42
	2	-1.3	0	15	0	1.70	0.33
	3	-1.3	0	24	0	3.56	0.27
	4	-0.65	0	6	0	1.03	0.37
	5	-0.65	0	15	0	2.09	0.31
	6	-0.65	0	24	0	3.40	0.27
	7	0	0	6	0	1.80	0.35
	8	0	0	15	0	2.44	0.31
	9	0	0	24	0	3.61	0.26
	10	0.65	0	6	0	2.36	0.31
	11	0.65	0	15	0	2.84	0.29
	12	0.65	0	24	0	3.85	0.27
	13	1.3	0	6	0	3.10	0.27
	14	1.3	0	15	0	3.58	0.27
	15	1.3	0	24	0	4.22	0.26
Static	16	-1.3	0	0	0	0.90	0.38
	17	-0.65	0	0	0	2.39	0.30
	18	0	0	0	0	3.89	0.26
	19	0.65	0	0	0	5.86	0.24
	20	1.3	0	0	0	7.39	0.25
In-depth motion	21	0	1.3	0	1	4.70	0.25
	22	0	1.3	0	2	5.35	0.24
	23	0	1.3	0	3	5.76	0.24
	24	0	2	0	1	4.82	0.25
	25	0	2	0	2	5.45	0.24
	26	0	2	0	3	6.14	0.24
	27	0	2.6	0	1	4.99	0.25
	28	0	2.6	0	2	5.95	0.24
	29	0	2.6	0	3	6.24	0.24
	30	-0.65	1.3	0	1	4.00	0.26
	31	-0.65	1.3	0	3	5.50	0.24
	32	-0.65	0.65	0	1	3.97	0.27
	33	-0.65	0.65	0	3	5.00	0.25
	34	0.65	1.3	0	1	5.34	0.24
	35	0.65	1.3	0	3	6.04	0.24
	36	0.65	0.65	0	3	5.99	0.24

10.3.2.5   Viewers

Viewer is another important influence factor on the experimental results as the gender, the age, or the viewing experience of the viewers would affect the experimental results. Viewer's information should be well documented and these factors should be balanced. For example, the gender and the age should be approximately uniformly distributed.

In this study, 42 viewers participated in Experiment 1. 21 are male, 21 are female. They are all non-experts in the domain of subjective experiments, image processing, or 3D. Their age ranged from 19 to 48 with an average age of 26.8. Ten experts in 3D perception, coding, quality assessment, and subjective experiments participated in the Experiment 2-a. Eight experts are male, two are female. Their ages ranged from 24 to 43 with an average age of 27. As the number of viewers is too small, to generate a reliable result, 5 of them conducted the test twice but on a different day. Thus, for each pair, there are 15 observations in total. 45 naive viewers participated in the Experiment 2-b. Twenty-one are male, 24 are female. They are all non-expert in subjective experiment, image processing, or 3D related field. Their ages ranged from 18 to 44 with an average age of 24.

All of the viewers have either normal or corrected-to-normal visual acuity. The visual acuity test was conducted with a Snellen Chart for both far and near vision. The Randot Stereo Test was applied for stereo vision acuity check, and Ishihara plates were used for color vision test. All of the viewers passed the pre-experiment vision check.

10.3.2.6   Assessment Method

The paired comparison method was used in this test. As there are 36 stimuli in Experiment 1, to reduce the number of comparisons, the ASD method was used [42]. In Experiment 2, as only 15 stimuli were considered, the FPC method was used. In Experiment 1, according to ASD method, 36 stimuli lead to a total of 180 pairs for each viewer. In Experiment 2-a, only 10 experts conducted the experiment. In order to produce more reliable results, each viewer conducted the test using the FPC (Full Pair Comparison) method and both presentation orders for each pair were considered, i.e., each stimulus pair will be shown twice but with different order, the stimulus which is shown firstly in one trial will be shown secondly in another trial. Thus, there were in total 210 pairs for each viewer in Experiment 2-a. In Experiment 2-b, a total of 105 pairs were presented to each viewer. The presentation order of the stimulus pair was different for odd numbered and even numbered viewers. For example, viewers with even numbers will watch stimulus *A* first, then stimulus *B*. For odd numbered viewers, this order is inverted. This is used to balance the presentation order. The presentation order of each stimulus pair in all experiments was randomly permuted for each viewer.

10.3.2.7 Procedures

The subjective experiments contained a training session and a test session. Five pairs of stimuli were shown in the training session. The viewers were asked not to stare at the moving object but to watch the whole stereoscopic sequence. Then, they should select the one which is more uncomfortable, concerning e.g., mental uneasiness. The viewers used two distinct keys to select one of the two stimuli in a pair for immediate visually check on the screen. There was a minimum duration for the display of each stimulus before making a decision by pressing a specific third button. The minimum duration is defined as either 5 s or the duration of a complete cycle of movement (the moving object went back to its start point) whichever was longer. During the training session, all questions of the viewers were answered. We ensured that after the training session, all of the viewers were familiar with the process and the task of this experiment.

In the main test session, 180 pairs were compared for Experiment 1, 210 pairs for Experiment 2-a, and 105 pairs for Experiment 2-b. As the duration of each test was different due to the number of pairs and individual differences of each viewer, and to avoid visual fatigue caused by long time watching affecting the experimental results, the Experiment 1 and Experiment 2-b were split into two sub-sessions. Each session contained half of the total number of stimulus pairs. There was a 10-min break between the two sub-sessions. For Experiment 2-a, the viewers were asked to have a 15 min break after each 30 min of the test.

10.3.3 Results of Experiment 1: Influence of Motion

As in this experiment the paired comparison methodology was used, the obtained results are paired comparison data for each pair. For better analysis, the Bradley–Terry (BT) model [43, 44] was used to convert the pair comparison data to psychophysical scale values for all stimuli. The BT scores and confidence intervals of all stimuli are shown in Table 10.3. For easier comparison, the lowest BT score is set to 0. The static condition can be considered as either a special case of the planar motion or in-depth motion, both with the motion velocity of 0. Thus, in this section, the static condition is analyzed in both conditions.

10.3.3.1 Planar motion and Static Conditions

The BT scores for the planar motion stimuli and static stimuli are shown in Fig. 10.6 where the static condition can be considered as a special case of the planar motion.

The experimental results on the planar motion stimuli showed that:

- Visual discomfort increases with the planar motion velocity;

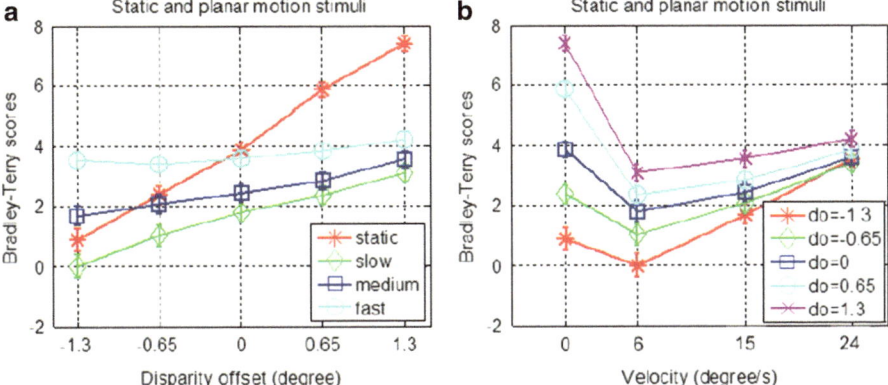

Fig. 10.6 The Bradley–Terry scores of the static and planar motion stimuli. (**a**) *Different lines* represent different velocity levels, where static, slow, medium, and fast represent 0, 6, 15, and 24°/ s. (**b**) *Different lines* represent different disparity offset levels. Bradley–Terry scores represent the degree of visual discomfort. Error bars are 95 % confidence intervals of the BT model fit

- The vergence–accommodation (VA) conflict might not significantly affect the visual discomfort. As shown in Fig. 10.6a, the visual discomfort neither reaches the minimum at the screen plane nor increases with the absolute value of disparity offset.
- The relative disparity r_o between the foreground and the background ($r_o = d_o + 1.4$ in this study) determines the visual discomfort, i.e., visual discomfort increases with the relative disparity.

A possible explanation of the influence of VA conflict and relative disparity on visual discomfort in our study might be the existence of the background. During the test, the viewers switched their attention between the background and the foreground, the larger this distance, the larger the change of VA conflict, which may lead to more visual discomfort.

For the static stimuli as shown in Fig. 10.6a, the visual discomfort increases with the relative disparity as well. Under the condition of small relative disparity, i.e., $r_o = 0.1°$ ($d_o = -1.3°$), the visual discomfort induced by static stimuli is less than the planar motion stimuli with medium or fast velocities. However, the gradient of the curve for the static case is steeper than the planar motion conditions. Thus, the visual discomfort increases faster with the disparity offset for the static stimuli than for the planar motion stimuli. By interpolating the static stimulus curve, when the static stimuli is very close to the background, i.e., with disparity offset close to $-1.4°$, the generated visual discomfort might be similar to the condition where the stimulus with similar disparity offset but slow planar motion. In the condition of disparity offset equals to zero degree ($r_o = 1.4°$), the static stimuli would generate similar visual discomfort as the fast planar motion stimuli. When the disparity offset is larger than 0°, the visualization of static objects seems to induce more visual discomfort than planar motion stimuli.

A different interpretation of these results is that when the relative disparity between the foreground and the background is increasing and the disparity offset becomes crossed, the planar motion seems to help to reduce visual discomfort when compared to the static condition. In our experiment, the viewers explained that when watching the planar motion stimuli, it was easier to fuse the Maltese cross compared to the static conditions, in particular when the disparity is crossed.

As shown in Fig. 10.6b, for the planar motion condition, there might be a minimum in the curve of visual discomfort that would be located at some velocity in between static and the slowest velocity that was included in this study.

10.3.3.2 In-Depth Motion and Static Conditions

The Bradley–Terry scores for in-depth motion stimuli are shown in Fig. 10.7. According to the results, we may draw the following conclusions:

- As shown in Fig. 10.7a, in general, disparity amplitude may not affect the visual discomfort significantly. For example, in the condition of $d_o = -0.65°$ and slow velocity, the visual discomfort induced by the stimulus with disparity amplitude of 0.65° is not significantly different from the stimulus with disparity amplitude of 1.3°. However, for the fast motion conditions, disparity amplitude may influence visual discomfort.
- Visual discomfort increases with the disparity offset as shown in Fig. 10.7a. This result is similar to the effect of relative disparity offset on planar motion stimuli, i.e., the relative disparity might be a main factor in this case.
- As shown in Fig. 10.7b, visual discomfort increased with the in-depth motion velocity.

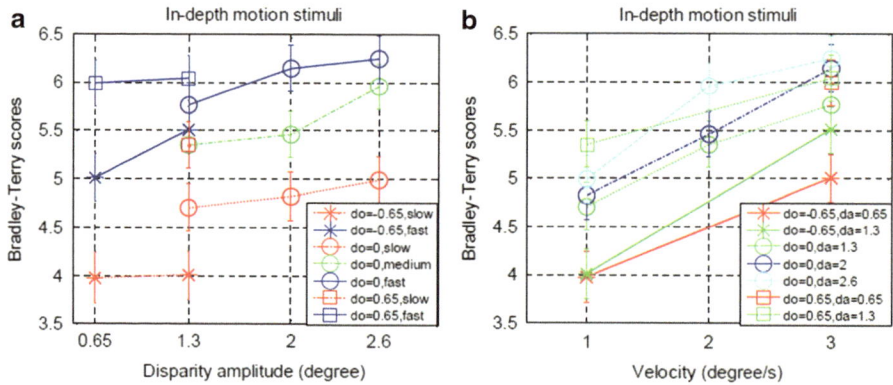

Fig. 10.7 The Bradley–Terry scores of the in-depth motion stimuli. (**a**) The x-axis represents the disparity amplitudes, *different lines* represent different disparity offsets (d_o) and velocities. (**b**) The x-axis represents disparity velocities, *different lines* represent different disparity offsets (d_o) and disparity amplitudes (d_a)

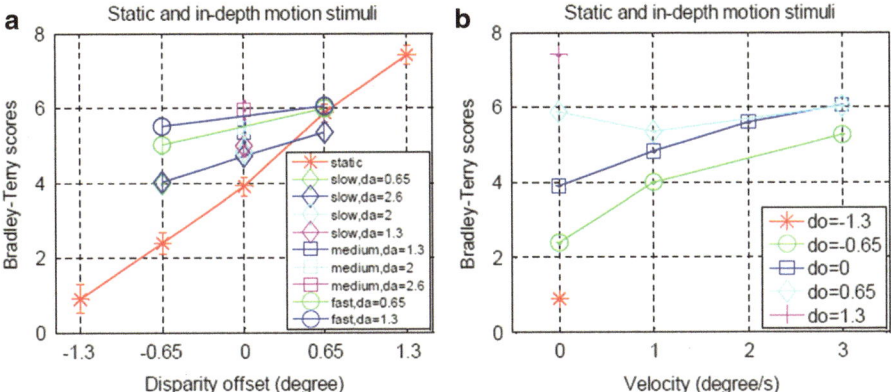

Fig. 10.8 The Bradley–Terry scores of the static and the in-depth motion stimuli. (**a**) The x-axis represents the disparity offsets, *different lines* represent different disparity amplitudes (d_a) and velocities. (**b**) The x-axis represents velocity. *Different lines* represent different disparity offset (r_o). Note that the BT score is the mean scores of the stimuli with same offset and velocity but different disparity amplitudes

The static condition can be considered as a special case of the in-depth motion as well. The BT scores of the static stimuli were compared with the in-depth motion stimuli which are shown in Fig. 10.8.

As shown in Fig. 10.8a, the gradient of the curve for in-depth motion is much flatter than for the static conditions. When the disparity offset is less than 0.65°, the in-depth motion will generate more visual discomfort than the static stimuli. For example, when compared with the static stimuli with the disparity of 0°, all the in-depth motion stimuli in our study generated more visual discomfort. However, when the relative disparity is larger than 2.05° ($d_o = 0.65°$), we may extrapolate that the visual discomfort induced by the static stimuli would be higher than the in-depth motion stimuli.

Considering the velocity, as shown in Fig. 10.8b, when the relative disparity is less than 2.05° ($d_o = 0.65°$), the visual discomfort increases with the velocity. However, if the relative disparity is larger than 2.05°, the static stimuli might generate more visual discomfort than the in-depth motion stimuli with slow velocity.

10.3.3.3   Discussions

In the current literature it is often mentioned that the motion in stereoscopic videos would induce more visual discomfort than static conditions. However, in this study, a counter-indication was found.

All three motion types showed that the relative disparity between the foreground and the background is a main factor in visual discomfort, i.e., visual discomfort increases with the relative disparity. The gradient of visual discomfort with relative disparity is highest for the static stimuli, followed by in-depth and then planar motion stimuli. This implies that static stimuli induce more visual discomfort when the relative disparity exceeds a certain value. This value is approximately 1.4° for planar motion and 2.05° for in-depth motion.

The gradient analysis also reveals that there is no "crossing point" between the planar motion and the in-depth motion in the positive three-quarters of the disparity space, i.e., from −0.65 to 1.3°. The in-depth motion stimuli are always more uncomfortable than the planar motion stimuli in this study. However, for the condition that the disparity offset is less than 0.65°, we may extrapolate that the slow in-depth motion stimuli might generate less visual discomfort than the fast planar motion stimuli. However, further studies are required.

10.3.4 Linear Regression Analysis: Toward an Objective Visual Discomfort Model

To investigate the influence factors of each motion type, multiple linear regression analysis is used in this study which attempts to model the relationship between two or more explanatory variables and a response variable by fitting a linear equation to observed data.

For the static situation, there is only one possible factor which is the relative angular disparity. For motion stimuli, the relative disparity offset, disparity amplitude, planar motion velocity, in-depth motion velocity, and their interactions are possible factors. The stepwise regression function in Matlab was used to select the most significant factors or remove the least significant factors [45]. The output of the stepwise regression includes the estimates of the coefficients for all potential factors, with confidence intervals, the statistics for each factor, and for the entire model. To avoid over-fitting of the model, the Leave-one-out Cross Validation (LOOCV) method was used for all possible models to find the model with the minimum averaged RMSE. The selected models are shown in Table 10.4.

Table 10.4 The linear regression analysis results for all stimuli

Motion type	Factor analysis			Model analysis
	Factor	Coefficient	p-value (t-test)	$D = \text{intercept} + \Sigma \text{ coefficient} \times \text{factor}$
Planar motion	r_o	1.45	0.0000	Intercept $= -1.11$
	v_p	0.18	0.0000	$R^2 = 0.98$, RMSE $= 0.17$
	$r_o \times v_p$	−0.04	0.0000	$F = 221.474$, p-value $= 4.10 \times 10^{-10}$
Static stimuli	r_o	2.53	0.0000	Intercept $= 0.54$
				$R^2 = 0.99$, RMSE $= 0.15$
				$F = 1176.37$, p-value $= 5.45 \times 10^{-5}$
In-depth motion	r_o	1.23	0.0000	Intercept $= 2.51$
	v_d	0.31	0.0000	$R^2 = 0.98$, RMSE $= 0.11$
	$d_a \times v_d$	0.45	0.0001	$F = 147.18$, p-value $= 1.8 \times 10^{-9}$
	$r_o \times d_a \times v_d$	−0.21	0.0031	

All the factors shown in Table 10.4 are statistical significant factors with p-value of the student's t-test < 0.05. The coefficient in the table is the coefficient of the corresponding factor in the linear model. The model analysis shows the linear model for each motion type. The R^2, RMSE, the F-statistic, and its p-value are provided as the evaluation results of this model. D represents visual discomfort score. It is shown that for the planar motion stimuli, the relative disparity, planar motion velocity, and their interaction term are important factors for visual discomfort. For the static stimuli, the relative disparity offset in this study shows its predominant effect. For the in-depth motion stimuli, the disparity amplitude is not a main factor. However, the interaction term of the disparity amplitude and the velocity, and the combination of the three factors (velocity, disparity amplitude, relative disparity offset) play an important role in determining visual discomfort.

According to the regression analysis results, an objective model for comparing visual discomfort of still stereoscopic images, planar motion stimuli, and in-depth motion stimuli is developed. All disparity and velocity values are measured in visual angular degree. Here we rewrite it as:

$$D = \begin{cases} 2.53r_o + 0.54 & \text{static} \\ 1.45r_o + 0.18v_p - 0.04r_ov_p - 1.11 & \text{planar} \\ 0.31r_o + 1.23v_d + (0.45 - 0.21r_o)d_av_d + 2.51 & \text{indepth} \end{cases} \tag{10.3}$$

The scatter plot of the objective and subjective results is shown in Fig. 10.9. The Pearson Linear Correlation Coefficient (PLCC), Spearman's Rank Correlation Coefficient (ROCC), and Root Mean Square Error (RMSE) are used to evaluate the correlation between the objective scores and the subjective results, and they are 0.9976, 0.9967, and 0.1198, respectively.

As this model is based on the paired comparison results, the D can be used to compare the degree of visual discomfort between the stimuli. The difference can be interpreted as the probability that one condition is preferred to another.

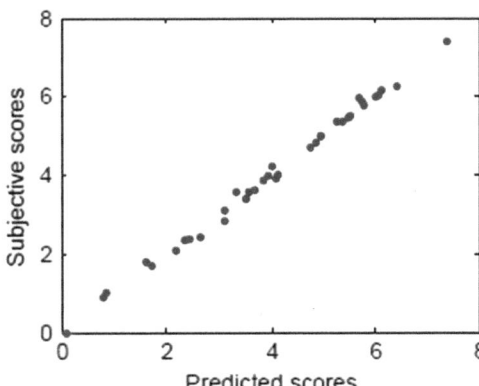

Fig. 10.9 The scatter plot of the predicted scores and the BT scores

10.3.5 Results of Experiment 2: Influence of Human Factors on Visual Discomfort

10.3.5.1 Comparison Between Experts and Naive Viewers

The BT scores of the Experiment 2 from experts and non-experts data are shown in Fig. 10.10. Both the experts and non-experts BT scores for the 15 planar motion stimuli provide the same conclusion as found in Sect. 10.3.3. The consistency of the experts and naive viewers' test results are: $CC = 0.9688$, $ROCC = 0.9357$, $RMSE = 0.2737$. The Barnard's exact test is applied on the raw pair comparison data of the experts and naive viewers results, and there are in total 21 pairs significantly different ($p < 0.05$), which corresponds to 20 % of the whole pairs. Thus, in general, the two experimental results are well correlated.

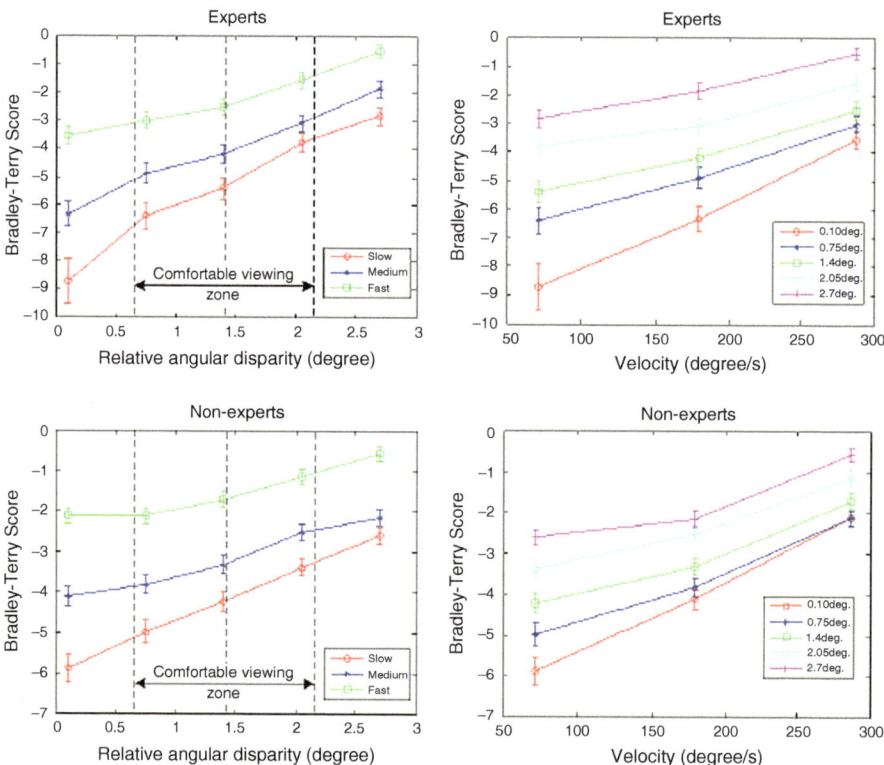

Fig. 10.10 BT scores for visual discomfort. The *top two figures* are experts' results. The *bottom two figures* are non-experts' results. The *different lines* in the *left figures* represent the different velocity levels. The *vertical two dashed lines* represent the *upper* and *lower limits* of the comfortable viewing zone, which are at 0.66 and 2.14°. The *dashed line* in the middle represents the position of screen plane. The *different lines* in the *right figures* represent the different relative angular disparity levels. The error bars are the 95 % confidence intervals of the BT model fit

10.3.5.2   Classification of Observers

When considering the influence from relative disparity and velocity, people may have different sensitivity on them. Thus, it may be interesting to classify them into different groups and analyze the different influences of relative disparity and velocity on different observers.

The relative disparity and the planar velocity are two factors that may induce visual discomfort in our study. Thus, the analysis which factor is dominant in determining the visual discomfort is conducted on each observer. There are two hypotheses in this analysis:

- Hypothesis 1: the relative disparity is predominant;
- Hypothesis 2: the velocity is predominant.

The methods to measure which factor is more predominant are based on the $p1$ and $p2$ values, where:

- $p1$: the proportion of each observer voting for the stimulus whose relative disparity is larger;
- $p2$: the proportion of voting for the stimulus whose velocity is faster than the other one.

Each observer's opinion on these two hypotheses can be reflected by $(p1, p2)$ which can be expressed by a point in a two-dimensional space. According to these points, the observers can be classified as different groups. In our study, the K-means clustering method was used. For better illustration, we define the term G-H1 (Group of Hypothesis 1) to represent the observer group who voted more according to Hypothesis 1, which means relative disparity is predominant in determining visual discomfort. A similar definition is used for G-H2. G-H12 is for the group who are equally sensitive to relative disparity and velocity, like the global subjective results. The clustering results are shown in Fig. 10.11. The BT scores for all stimulus generated by each observer cluster are shown in Fig. 10.12.

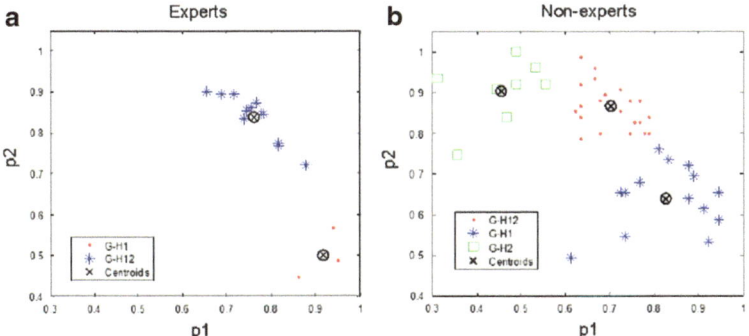

Fig. 10.11 The clustering results for experts and non-experts observers. x-Axis represents the agreement on "relative disparity is the predominant factor" and y-axis represents the agreement on "velocity is the predominant factor"

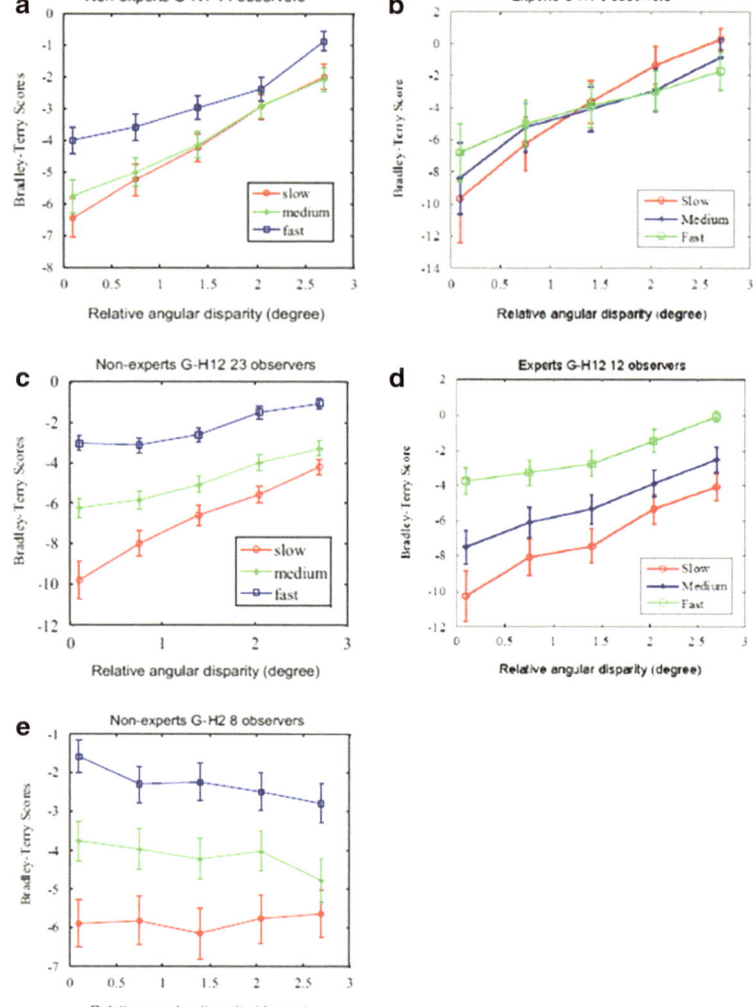

Fig. 10.12 BT scores of the different classes of the viewers. Naive viewers' results are in the first column. Experts' results are in the second column. The rows represent group G-H1, G-H12, and G-H2 respectively

According to Fig. 10.12 it is shown that most of the viewers agree with the global subjective experiment results, i.e., visual discomfort increases with the motion velocity and relative disparity. There are a small number of viewers who have a completely different sensitivity on relative disparity and motion velocity. It should be noted that for the results of Experts group G-H1 the number of viewers is very small and the results may not be reliable as there are only three observations for each pair.

Conclusions

This chapter provided important insight concerning the influence factors of visual discomfort and their measurement. Technological constraints in the current stereoscopic production chain inhibit the complete elimination of visual discomfort but careful control of technical parameters allows for reducing the negative effects. It was shown that parameters may need to be adjusted to individual observer groups.

As visual discomfort can only be subjectively measured, special care needs to be taken with respect to the experimental setup and the statistical data analysis. The chapter focused on a complete example study evaluating the influence of planar and in-depth motion at different velocities, starting with the precise description of the viewing setup and finishing with a model of visual discomfort, and analyzing its limits with respect to individual observer groups. Several important conclusions can be drawn from this study: Slow motion may be more comfortable at the same disparity than static scenario, in-depth motion is more uncomfortable than planar motion, the existence and position of a fixed background plays an important role in 3D psychovisual stimulus studies, bimodal distributions of observer responses should be expected. A general model of spatiotemporal 3D motion was presented that fits most of the participating population.

References

1. Devroede G (1992) Constipation–a sign of a disease to be treated surgically, or a symptom to be deciphered as nonverbal communication? J Clin Gastroenterol 15(3):189
2. King LS (1982) Medical thinking: a historical preface. Princeton University Press Princeton, NJ
3. Natelson BH (1998) Facing and fighting fatigue: a practical approach. Yale University Press, New Haven
4. Urvoy M, Barkowsky M, Li J, Le Callet P (2013) Visual comfort and fatigue in stereoscopy. In: 3D video: from capture to diffusion, Chapter 16, Wiley-ISTE, Laurent Lucas and Céline Loscos and Yannick Rémion, October 2013, pp 309–329
5. Hoffman DM, Girshick AR, Akeley K, Banks MS (2008) Vergence–Accommodation conflicts hinder visual performance and cause visual fatigue. J Vis 8(3):1–30
6. Kim J, Shibata T, Hoffman D, Banks M (2011) Assessing vergence accommodation conflict as a source of discomfort in stereo displays. J Vis 11(11):324–324
7. Yano S, Ide S, Mitsuhashi T, Thwaites H (2002) A study of visual fatigue and visual comfort for 3D HDTV/HDTV images. Displays 23(4):191–201
8. Chen W, Fournier J, Barkowsky M, Le Callet P (2013) New requirements of subjective video quality assessment methodologies for 3DTV. In: International workshop on video processing and quality metrics
9. Yano S, Emoto M, Mitsuhashi T (2004) Two factors in visual fatigue caused by stereoscopic HDTV images. Displays 25(4):141–150
10. Kuze J, Ukai K (2008) Subjective evaluation of visual fatigue caused by motion images. Displays 29(2):159–166

11. Speranza F, Tam WJ, Renaud R, Hur N (2006) Effect of disparity and motion on visual comfort of stereoscopic images. Proc SPIE Stereoscop Displays Virtual Reality Syst 6055:94–103
12. Tam WJ, Speranza F, Vazquez C, Renaud R, Hur N (2012) Visual comfort: stereoscopic objects moving in the horizontal and mid-sagittal planes. In: Proc. SPIE 8288, Stereoscopic Displays and Applications XXIII, 828813
13. Mendiburu B (2009) 3D movie making: stereoscopic digital cinema from script to screen. Focal Press, Burlington
14. Tam WJ, Speranza F, Yano S, Shimono K, Ono H (2011) Stereoscopic 3D-TV: visual comfort. IEEE Trans Broadcast 57(2):335–346
15. Ide S, Yamanoue H, Okui M, Okano F, Bitou M, Terashima N (2002) Parallax distribution for ease of viewing in stereoscopic hdtv. In: Electronic imaging 2002, International Society for Optics and Photonics, pp 38–45
16. Nojiri Y, Yamanoue H, Hanazato A, Okano F (2003) Measurement of parallax distribution, and its application to the analysis of visual comfort for stereoscopic hdtv. Proc SPIE 5006:195–205
17. Nojiri Y, Yamanoue H, Ide S, Yano S, Okana F (2006) Parallax distribution and visual comfort on stereoscopic hdtv. In: Proc. IBC, pp 373–380
18. Kooi FL, Toet A (2004) Visual comfort of binocular and 3D displays. Displays 25(2–3):99–108
19. Pastoor S (1995) Human factors of 3d imaging: results of recent research at heinrich-hertz-institut berlin. Proc IDW 95:69–72
20. Tyler CW, Likova LT, Atanassov K, Ramachandra V, Goma S (2012) 3D discomfort from vertical and torsional disparities in natural images. In: Proc. 17th SPIE Human Vision Electron. Imaging, 8291:82910Q-1
21. Kim D, Sohn K (2011) Visual fatigue prediction for stereoscopic image. IEEE Trans Circuits Syst Video Technol 21(2):231–236
22. Woods AJ, Docherty T, Koch R (1993) Image distortions in stereoscopic video systems. In: IS&T/SPIE's Symposium on Electronic Imaging: Science and Technology, International Society for Optics and Photonics, pp 36–48
23. Ijsselsteijn WA, de Ridder H, Vliegen J (2000) Subjective evaluation of stereoscopic images: effects of camera parameters and display duration. IEEE Trans Circuits Syst Video Technol 10 (2):225–233
24. Yamanoue H, Nagayama M, Bitou M, Tanada J, Motoki T, Mituhashi T, Hatori M (1998) Tolerance for geometrical distortions between l/r images in 3D-HDTV. Syst Comput Japan 29 (5):37–48
25. Jang TY (2012) A study of visual comfort of 3d crosstalk and binocular colorrivalry thresholds for stereoscopic displays. Master's thesis, National Taiwan University of Science and Technology
26. Seuntiëns PJH, Meesters LMJ, Ijsselsteijn WA (2005) Perceptual attributes of crosstalk in 3D images. Displays 26(4–5):177–183
27. Chen L, Tu Y, Liu W, Li Q, Teunissen K, Heynderickx I (2008) 73.4: Investigation of Crosstalk in a 2-View 3D Display. In: SID symposium digest of technical papers, vol. 39, Wiley Online Library, pp 1138–1141
28. Hanazato A, Okui M, Yuyama I (200) Subjective evaluation of cross talk disturbance in stereoscopic displays. In: SDI 20th Int. Display Res. Conf, pp 288–291
29. Wang L, Tu Y, Chen L, Zhang P, Teunissen K, Heynderickx I (2010) Crosstalk acceptability in natural still images for different (auto) stereoscopic display technologies. J Soc Inform Display 18(6):405–414
30. Harris JM, McKee SP, Watamaniuk SNJ (1998) Visual search for motion in-depth: stereomotion does not pop out from disparity noise. Nat Neurosci 1(2):165–168
31. Cho S-H, Kang H-B (2012) An assessment of visual discomfort caused by motion-in-depth in stereoscopic 3D video. In: Proceedings of the British Machine Vision Conference, BMVA Press, pp 65.1–65.10

32. Harris LR, Jenkin MRM (2011) Vision in 3D environments. Cambridge University Press, Cambridge
33. Lee S, Jung YJ, Sohn H, Ro YM, Park HW (2011) Visual discomfort induced by fast salient object motion in stereoscopic video. Proc SPIE 7863:786305
34. Li J, Barkowsky M, Wang J, Le Callet P (2011) Study on visual discomfort induced by stimulus movement at fixed depth on stereoscopic displays using shutter glasses. In: 17th International Conference on Digital Signal Processing (DSP), IEEE, pp 1–8
35. Li J, Barkowsky M, Le Callet P (2011) The influence of relative disparity and planar motion velocity on visual discomfort of stereoscopic videos. In: International workshop on quality of multimedia experience, pp 155–160
36. ITU-R BT.500-13 (2012) Methodology for the subjective assessment of the quality of television pictures. International Telecommunication Union, Geneva, Switzerland
37. Brainard DH (1997) The psychophysics toolbox. Spatial Vision 10(4):433–436
38. Pelli DG (1997) The VideoToolbox software for visual psychophysics: transforming numbers into movies. Spatial Vision 10(4):437–442
39. Fukushima T, Torii M, Ukai K, Wolffsohn JS, Gilmartin B (2009) The relationship between CA/C ratio and individual differences in dynamic accommodative responses while viewing stereoscopic images. J Vis 9(13), article 21
40. McLin LN, Schor CM (1988) Voluntary effort as a stimulus to accommodation and vergence. Investigat Ophthalmol Vis Sci 29(11):1739–1746
41. OHare L, Hibbard PB (2011) Spatial frequency and visual discomfort. Vis Res 51(15):1767–1777
42. Li J, Barkowsky M, Le Callet P (2013) Boosting Paired Comparison methodology in measuring visual discomfort of 3DTV: performances of three different designs. In: Proceedings of the SPIE electronic imaging, stereoscopic displays and applications
43. Bradley RA (1984) 14 paired comparisons: Some basic procedures and examples. Handbook Stat 4:299–326
44. Bradley RA, Terry ME (1952) Rank analysis of incomplete block designs: I. The method of paired comparisons. Biometrika 39(3/4):324–345
45. Draper NR, Hy S, Pownell E (1981) Applied regression analysis, vol 3, 2nd edn. Wiley, New York

Chapter 11
3D Sound Reproduction by Wave Field Synthesis

Hyun Lim

Abstract In this chapter, we focus on the spatial audio technique implemented for 3D multimedia content including spatial audio and video. The spatial audio rendering method based on wave field synthesis is particularly useful for applications where multiple listeners experience a true spatial soundscape while being free to move around without losing spatial sound properties. The approach can be considered as a general solution to the static listening restriction imposed by conventional methods, which rely on an accurate sound reproduction within a sweet spot only. While covering the majority of the listening area, the approach based on wave field synthesis can create a variety of virtual audio objects at target positions with very high accuracy. An accurate spatial impression could be achieved by wave field synthesis with multiple simultaneous audible depth cues improving localisation accuracy over single object rendering. The current difficulties and practical limitations of the method are also discussed and clarified in this chapter.

11.1 Introduction

An outline of the concept is given here to help the understanding the background of the system design. Wave Field Synthesis (WFS) is a sound field reconstruction methodology for rendering true spatial sound within a given space. WFS is a practical application of the Huygens' Principle [1]. According to the principle when a wave is generated by a point source, all the points on its wave front can be replaced with a point source array producing of spherical secondary wavelets [2]. In WFS the point source array can be replaced with a continuous active sound source array and the wave field within a volume can be synthesised. The main feature of the methodology discussed in this chapter is its ability to automatically

H. Lim (✉)
Institute for Digital Technologies, Loughborough University in London,
Queen Elizabeth Olympic Park, London E20 3ZH, UK
e-mail: H.Lim@lboro.ac.uk

© Springer Science+Business Media New York 2015
A. Kondoz, T. Dagiuklas (eds.), *Novel 3D Media Technologies*,
DOI 10.1007/978-1-4939-2026-6_11

reproduce spatial sound equivalent to one generated at the original object position. This method of sound field reconstruction by WFS can be implemented with multiple input audio signals using audio object information for accurate replication.

An application scenario that could be rendered by the WFS-based spatial audio system would typically be where users are in a room and enjoying other multimedia applications, e.g. 3D multi-view video [3], together with spatial audio [4]. In recent years streaming 3D multi-view video and its associated spatial audio to multiple users simultaneously have been a highly challenging area in immersive audio–visual applications. Suppose that a broadcaster is delivering this scenario using high quality 3D multi-view content through the Internet with a synchronised spatial audio service. The proposed approach aims to implement an optimised solution in a listening space corresponding to a video screens viewing area. The audio signal is processed appropriately for the quality layers of colour and depth map of 3D multi-view video to the rendered sound field.

At the receiving end 3D content rendering would require multi-channel audio equipment and extended displays. In practical implementation of an audio–visual system with WFS a big shortfall is a bulky loudspeaker array, since it unnecessarily attracts listeners' visual attention and also is not economical. Recently several simplified WFS applications have been developed, which were implemented with few loudspeakers [5, 6].

We can suppose such a typical case that a video stream synchronized with audio objects is delivered to a display screen mounted in front of users. To achieve a more feasible design the listening angle need only be valid in the corresponding viewing angle of the primary display. Taking this into account, in practice, a loudspeaker array does not necessarily need to completely surround the entire domain, instead it can cover the side facing the viewer and run parallel to the main visualisation medium.

At the rendering side, spatial information is applied to create the mathematical processes used to drive the reproduction of the audio object's signal. The processing is done such that the object sounds as if it were at the specified true position and propagating along the true corresponding direction. Attributes such as sound level, position and orientation are therefore essential in this auditory scene reconstruction. The audio rendering system operates as a mixer which interprets the scene information and reproduces the sound field through the given number of transducer feeds, e.g. multi-channel loudspeakers, as mixtures of many synthesised objects. The audio–rendering application should be optimised appropriately for the acoustic environment of a listening space.

11.2 Theoretical Formulation

According to the Huygens' principle, if a wave generated by a point source, all the points on the wave front can be replaced with another point source array for the production of spherical secondary wavelets [2]. With a continuous secondary active sound source array, wave fronts within a volume can be synthesised. Kirchhoff and Helmholtz formulated the sound pressure P as an integral equation, Eq. (11.1) based on the Huygens' principle [7].

$$P(\vec{r}, \omega) = \frac{1}{4\pi} \oint_S \left[\underbrace{P(\vec{r}_s, \omega) \frac{\partial}{\partial n} \left(\frac{e^{-jk|\vec{r}-\vec{r}_s|}}{|\vec{r} - \vec{r}_s|} \right)}_{dipoles} - \underbrace{\frac{\partial P(\vec{r}_s, \omega)}{\partial n} \frac{e^{-jk|\vec{r}-\vec{r}_s|}}{|\vec{r} - \vec{r}_s|}}_{monopoles} \right] dS, \quad (11.1)$$

where k is a wave number, ω is the angular frequency of a wave, c is the speed of sound. $\vec{r}$ denotes a position vector of a listening position inside the domain surrounded by an arbitrary surface S. $\vec{r}_s$ is a position vector on the surface S. n is the internal normal on S. In reproduction of the sound field at $\vec{r}$, the former out of the two main terms on the right-hand side of Eq. (11.1) can be implemented as acoustic dipoles and the latter as monopoles situated on S. For practical implementation of the Eq. (11.1) we can choose only the monopole source term in the Kirchhoff–Helmholtz integral via modification of Green's function [8]. Theoretically Eq. (11.1) allows us to implement the WFS system with various combinations of a monopole and dipole. However a monopole is generally chosen as a secondary source for WFS in realisation due to the simple construction, more reliable controllability and relatively smaller in size than a dipole.

Usually known as Rayleigh I and Rayleigh II integral Eqs. (11.2) and (11.3) describe solutions for monopoles and dipoles respectively [9].

$$P(\vec{r}, \omega) = \rho_0 c \frac{jk}{2\pi} \iint_S \left(V_n(\vec{r}_s, \omega) \frac{e^{-jk|\vec{r}-\vec{r}_s|}}{|\vec{r} - \vec{r}_s|} \right) dS, \quad (11.2)$$

and

$$P(\vec{r}, \omega) = \frac{jk}{2\pi} \iint_S \left(P(\vec{r}_s, \omega) \frac{1 + jk|\vec{r} - \vec{r}_s|}{|\vec{r} - \vec{r}_s|} cos\varphi \frac{e^{-jk|\vec{r}-\vec{r}_s|}}{|\vec{r} - \vec{r}_s|} \right) dS, \quad (11.3)$$

where ρ_0 denotes the air density, and V_n the velocity normal on the boundary S surrounding the domain, c the speed of sound. φ denotes the relative angle between the vector from a position on S to the reference position L and the normal vector V_n in Fig. 11.1.

Equations (11.2) and (11.3) satisfy their solutions through the entire volume inside S. However, in practical approach the valid solutions for WFS can be found only on a horizontal plane [1, 10]. In that case, the surface integral can be reduced to a line integral [11]. In implementation of the solutions a continuous distribution of secondary sound sources, i.e. ideally a line source, has been supposed to be used. The practical limitation requires the equations described in discrete form. The

Fig. 11.1 Problem sketch

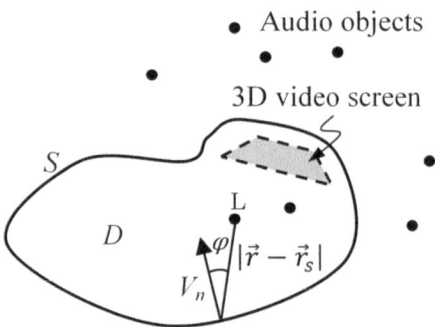

secondary source can be implemented with an array of number of discrete loudspeakers.

As the result, Eq. (11.1) can be simplified into the discrete form of Rayleigh I integral as shown in Eq. (11.4) as the following:

$$P(\vec{r}, \omega) = \frac{j\rho_0\omega}{2\pi} \sum_{m=1}^{\infty} \left[V_m(\vec{r}_m, \omega) \frac{e^{-jk|\vec{r}-\vec{r}_m|}}{|\vec{r}-\vec{r}_m|} \right] \Delta x. \tag{11.4}$$

The driving function for the m-th driver in the discrete control source array is obtained from (11.4) and is shown in Eq. (11.5).

$$Q_m(\vec{r}, \omega) = A_m(\vec{r}, \omega) W(\omega) \frac{e^{-jk|\vec{r}-\vec{r}_m|}}{|\vec{r}-\vec{r}_m|}, \tag{11.5}$$

where A_m denotes a weighting factor of the m-th driver, $W(\omega)$ is the frequency spectrum of an audio signal. The driving function (11.5) is a modified interpretation of the Rayleigh integral equation (11.2) describing the solution for reconstruction of a wave field surrounded by only monopoles. However, (11.5) is valid only when a virtual source is supposed to be situated outside S (see Fig. 11.1).

The driving function (11.5) can be modified further so that the solution is valid for reconstruction of a wave field both inside and outside S, in other words everywhere except on the boundary S in Fig. 11.1 [12]:

$$Q_m(\vec{r}, \omega) = A_m(\vec{r}, \omega) W(\omega) j^{\theta(\vec{r})} \frac{e^{(-1)^{\theta(\vec{r})+1} jk|\vec{r}-\vec{r}_m|}}{|\vec{r}-\vec{r}_m|}. \tag{11.6}$$

Here A_m denotes a weighting factor of the m-th driver, $W(\omega)$ is the frequency spectrum of an audio signal and $\theta(\vec{r})$ is an indicator function. $\theta(\vec{r})$ has the value 1 for all $\vec{r} \in D$, i.e. a focused source; otherwise 0, where D is a problem domain and S is the boundary of D. It is important to emphasise that the driving function shown in (11.6) is valid for a source generated both inside and outside the domain D. A key advantage in this approach over earlier studies on methodologies based on WFS [13, 14] is that it requires only one equation for both focused and non-focused sources. The single general equation is very useful for real-time applications related with both focused and non-focused sources. For example when an audio object is continuously moving from outside a domain, including movement through the loudspeaker array which corresponds to the boundary S in Fig. 11.1.

In order to validate the solution provided by (11.6), we analyse some typical results obtained in the following numerical simulations. Figure 11.2 shows results taken from an array of 16 loudspeakers mounted on one side of a given space. The problem domain where the solution is valid is defined in anechoic conditions in this simulation. The geometry can be given to reflect the actual physical boundary conditions of the studio where the actual WFS system takes place. When the solution of (11.6) is applied to the space, the combination of all the individual wave fronts produces a unified wave field.

In the simulation a single point source is virtually situated outside the listening room $(\theta(\vec{r}) = 0)$ in Fig. 11.2a, and also inside $(\theta(\vec{r}) = 1)$ in Fig. 11.2b. A display screen can be mounted in front of the loudspeaker array (c.f. Fig. 11.1) and is assumed acoustically transparent. Even though this is generated by a reduced number of drivers, instead of a contentious ideal line source, the reproduced wave field is equivalent to the original sound sources wave front in the most of the given domain in Fig. 11.2. Near the boundary, e.g. on the left in Fig. 11.2a, some artefacts can occur. However the artefacts are not uncommon in many other multi-channel surround audio techniques, such as ambisonics, 5.1 and 7.1 channel audio [15]. This can happen largely due to truncated secondary source array which can cause aliasing tails on the reconstructed wave front's truncated regions [9, 10, 15, 16]. Overall, the output results show wave field validity within the viewing angle, as indicated in Fig. 11.2b. It confirms that the general WFS solution based on the driving function in (11.6) can be used for both cases using sources created outside (Fig. 11.2a) and also inside (Fig. 11.2b) the domain. The general solution (11.6) is applicable in 2D and also in general 3D cases.

11.3 Implementation of WFS

11.3.1 General System Design

In implementation of WFS, ideally, the entire listening space should be surrounded by a continuous line-source. However, in practice the secondary source can be replaced with a series of discrete drivers as shown in Fig. 11.3 [9].

The control PC where the WFS algorithm is implemented is equipped with multi-channel audio interface cards, e.g. HDSP MADI interface card [17]. From here the signals are then fed via coaxial cables into AD/DA converters. In a studio a

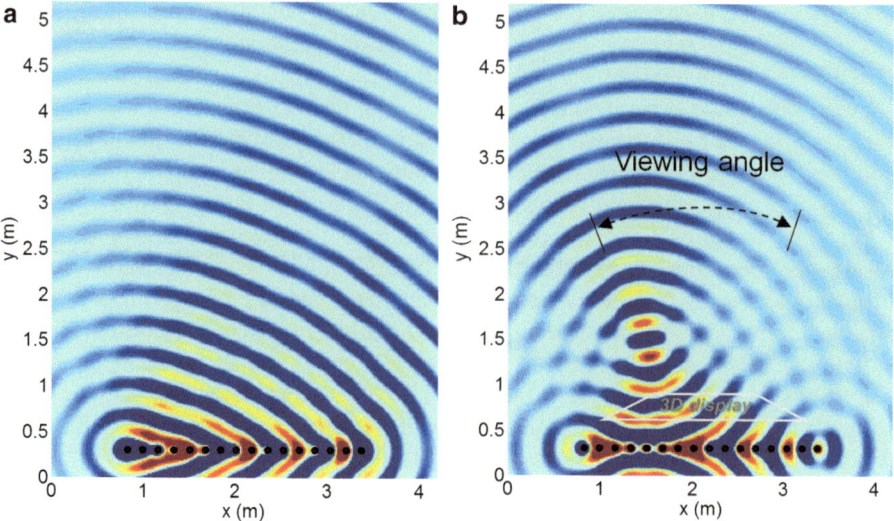

Fig. 11.2 Sound pressure distribution when (**a**) a virtual source is created at $(0, -4)$, and (**b**) a focused source at $(1.5, 1.5)$, 1 kHz

series of drivers is mounted and used as secondary acoustic monopole sources for sound field reproduction. A state-of-the-art standard audio format can be used for storage and transmission of an input audio scene, such as the Spatial Sound Description Interchange Format (SpatDIF) [18]. The signal for each channel is defined by special WFS software including the driving function (11.6), which computes the gain values and delay times of each loudspeaker in the array. These are needed to synthesise the virtual audio objects predefined by the input scenario. The data loaded from the input consists of the audio signal, object information, the sampling data, and so on. The object information includes the spatial position vector of each virtual object and in addition the geometry of the loudspeaker array is given in advance. According to further requirements, the function of the WFS system can be enhanced to enable moving sources [19] and focused sources, i.e. virtual sources inside the domain [13].

11.3.2   Hardware Design

Hardware modules for WFS are illustrated in Fig. 11.4. It is supposed that the measurement procedure measures beforehand the audio signals and additional objective information at the location of the original sources. The Audio data loaded from input is fed into the WFS system through ADAT[1]-audio converter first. The

---

[1] Alesis Digital Audio Tape protocol.

Fig. 11.3 An example of a loudspeaker array arrangement in a studio

Fig. 11.4 Block diagram
of main hardware including
DSP apparatus

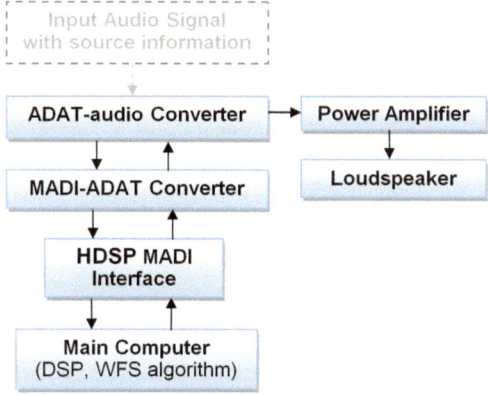

converted data are sent to ADAT–MADI[2] converter. After all these conversions, the resulting data are stored in a computer through MADI card and can be used for digital signal processing (DSP). MADI helps to enable easier communication to the WFS software installed in a computer without having to deal with raw audio data. The data received in the computer is then incorporated into the WFS algorithm together with the calibrated loudspeaker transfer functions and directivity to generate the desired sound signals. The data convolved with filter functions is saved as phase-synchronous files which can be played back using any multi-channel compatible audio editor. In the system, after all the filtering process the signals are led to the ADAT matrix which splits and transmits them to each channel. After being played back by the computer, the separate audio signals are converted to MADI and sent to the MADI–ADAT converter via RME[3] HDSP[4] sound cards with 64 outputs each in MADI format and then ADAT-audio converter successively. The sound generation system consists of loudspeaker arrays and power amplifiers. Each driver

---

[2] MADI: Multi-Channel Audio Digital Interface.

[3] RME Corporation.

[4] Hammerfall DSP.

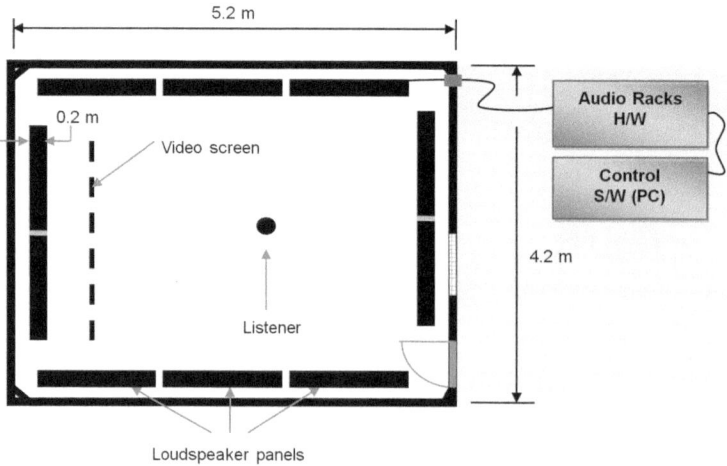

Fig. 11.5 Arrangement of loudspeaker panels for WFS in a studio

is designed with an enclosure. The optimised centre to centre distance between two neighbouring control sources is about 18–25 cm in many practical WFS applications. In addition, sub-woofers can also be used for low frequency compensation.

A possible plan view of the WFS system in a studio is illustrated in Fig. 11.5. The reverberation effects in the WFS algorithm can be not so much significant and rather acceptable for the system as the interior walls of the studio are covered by absorptive acoustic panels having an absorption coefficient $\alpha = 0.3$–0.4. α varies with the frequency of sound. The original approaches to WFS require anechoic environment to implement an ideal global sound field reconstruction [9]. The loudspeaker panels stand close to each side wall of the studio inside. The total number of loudspeakers used as secondary sources is generally proportional to the size of a listening room.

All the loudspeaker panels are controlled by the PC and hardware control modules situated outside the studio to avoid any unnecessary acoustic obstacle as shown in Fig. 11.5. The system requires number of audio channels to control all the loudspeakers individually. The control PC is equipped with MADI interface cards. An up-to-date MADI interface card, e.g. RME HDSP MADI, is capable of outputting 64 channels of audio in MADI format [17]. The MADI signals are then fed via coax into AD/DA converters which convert the signals into multi-channel audio outputs. The control software includes a series of multi-channel audio APIs as libraries that can be accessed easily from programming languages, such as C++, making it easy to implement highly complicated audio DSP techniques and WFS algorithms.

11.3.3   Software Design

The method of immersive sound field reconstruction can be implemented more effectively with object-based audio signals. Object-based spatial audio is a concept of delivering audio which operates by rendering each audio object that appears in the auditory scene [12]. In WFS, the auditory scene is created by rendering the sound field within a given listening area corresponding to the captured scene. The rendering module is responsible for distributing the sound objects within the generated sound field according to the scene description parameters.

Each input signal to a driver is defined by WFS algorithm which computes signals for the predefined virtual audio objects. The main process of DSP is illustrated in Fig. 11.6. The data loaded from the input consists of the audio signal, source information, sampling frequency, and so on. The source information includes the spatial position vector of each virtual sound source. In addition, the distance from the loudspeaker array to the listening position, the number of drivers and a gap between them are given in advance.

The first step of the algorithm is the calculation of the distance between each driver and the virtual source. The driving function calculates the time for each sample that the sound front takes from the virtual source arriving to each driver, based on the input data. Then the convolution of the output driving signal is performed with the impulse response of each driver. The signal after the convolution is described in the frequency domain through the FFT process so that the directivity can be included. The directivity of each driver according to the frequency and angle of emission is measured before the WFS algorithm starts. Generally, it can be measured in advance by the calibration process. The frequency contents are transformed back into the time domain via the inverse FFT process. The output vectors can be saved as audio files.

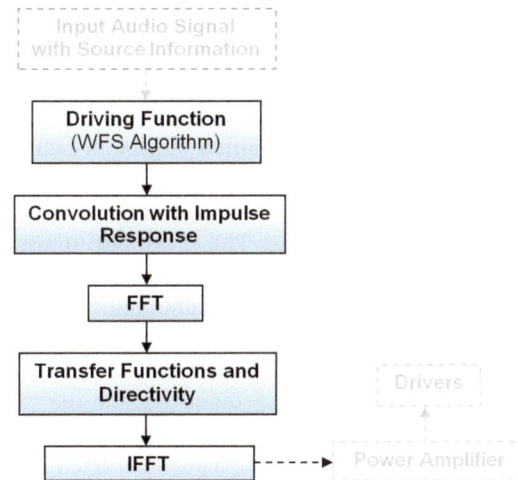

Fig. 11.6 Block diagram of main software

11.3.4   Virtual Source Localisation

The physical properties of the output sound field generated by an audio rendering system can directly be measured by using precise measuring microphones. This is done to objectively evaluate the accuracy of the acoustic sources being synthesised by (11.6). When the solution (11.6) based on the WFS theory is applied to the space, the combination of all the output wave fronts generated by the drivers produces a wave front equivalent to the original virtual one in a given domain. The result shows that the solution Eq. (11.6) which is used for a source created outside the domain is also effective and can be extended to the cases with a focused source. The position of a virtual source is experimentally determined by measuring the direction of a wave front's propagation. In order to identify the position of an acoustic source created by the rendering system, the perpendicular direction to a wave front is measured using the two microphone setup mounted at predefined reference positions. The acoustic measurement technique determines the maximum pressure gradient at the measuring point.

Figure 11.7 shows a representative example of the experimental results measuring the difference in position between the intended object and the rendered virtual object, where + denotes the reference position of the intended object. The measurements are taken from a studio with moderately absorptive walls to limit reverberations. The absorption coefficient $\alpha = 0.3$–0.4, varying with the frequency. The measuring system is composed of a combination of B&K half-inch free-field measuring microphones [20]. The phase difference between a pair of the microphones was measured using a B&K PULSE Sound & Vibration analyser [21]. An off-centre localisation condition is deliberately chosen in this particular experiment, shown as reference object position + in Fig. 11.7. This is chosen so that the result can present the higher limitation of the directional errors. Common artefacts by any WFS-based systems can usually be found near the ends of a truncated loudspeaker array [9, 10, 15, 16]. When the WFS system is switched on, the main propagating direction of the wave front is measured at each pair of multiple receiver positions, e.g., A_1 and B_1, A_2 and B_2, . . ., and A_N and B_N, where N is the total number of the measurements. The directional errors $\Delta\varphi$ are indicated in Fig. 11.7 and can be generally below $7°$ in directional accuracy [12]. Causes of the errors in the measurement can be classified largely in two groups, one related to position, and the other to time. Firstly we consider position, the physical size of the measuring devices and the gap between two neighbouring loudspeaker drivers cause aliasing errors while monitoring. The maximum uncertainties at which these errors take place correspond to about $10°$. In addition, time delay and extra phase errors can occur in DSP and measuring equipment. Therefore, the resulting errors are not uncommon in real environments. In the context of human hearing, the errors are acceptable as the directional error is within $10°$ limit [22].

Some artefacts can be found outside the viewing angle which is indicated in Fig. 11.7. This is due to the fact that the limited length of the secondary source array can cause discontinuity of a reconstructed wave front at each end. The spatial

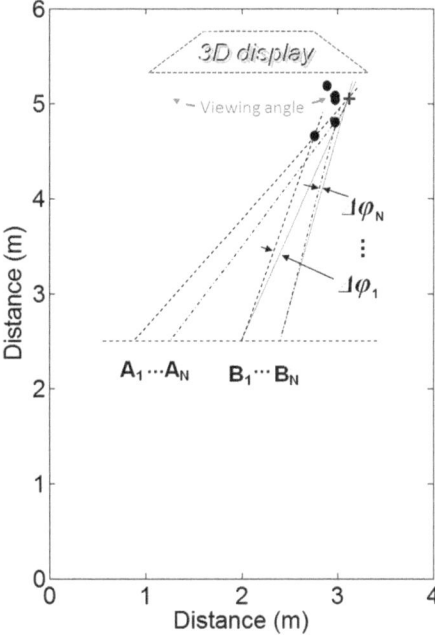

Fig. 11.7 Virtual source positions, (*plus symbol*): originally intended, (*bullet symbol*): measured at different positions, A_N and B_N

aliasing frequency can change at each listening position. The spatial aliasing may cause extra artefacts to the listener, e.g. colouration artefacts. Overall, the typical results shown as examples are mostly valid within the view angle.

In addition to the objective evaluation method, subjective evaluation methods are also available for audio rendering system validation as shown in earlier other works [23, 24]. The subjective evaluation can follow the ITU-R BS.1534-1 guidelines [25]. During evaluation the participants are asked to move around the whole room including outside the viewing angle. Subjects record the virtual sources origin and depth using a visual place marker and questionnaire.

The subjective evaluation was conducted in the studio illustrated in Fig. 11.5. Considering previous listening experiences, the listening panel was chosen for the test such as (a) Electronic engineer and recording musician, (b) Electronic engineer, (c) Audio/visual engineer with experience in spatial audio, (d) Broadcast and media professional, (e) DJ and musician. In Table 11.1 it can be seen that the participants accurately localised the audio objects origins over the majority of the listening space. It is also shown that non-focused sources (Table 11.1a) can be reproduced more accurately than focused (Table 11.1b), especially for the perception of depth.

It is found that the destructive acoustic errors of the focused sound source can be reduced when the listeners are within the calculated viewing angle. Another finding shows that the sound field generated from multiple virtual audio objects improve perceptual sensitivity, with a higher directional accuracy by approximately 6–14 %. This improvement can be explained using the acoustic side effects produced when

Table 11.1 Overall perceptual localization accuracy for (a) non-focused and (b) focused sources when single and multiple virtual audio objects are created

	Accuracy (%)	Deviation (%)
(a) Non-focused sources		
Multiple source	94	3.5
Single source	88.3	6.8
(b) Focused sources		
Multiple source	84	6
Single source	70	12

multiple broadband sound fields mix together. The side effect is called sound masking and can help reduce undesirable acoustic errors in a room while also increasing the ability to perceive depth and origin [26]. The sound masking effects can restore a more natural 3D audio environment.

Conclusions

The feasibility of an immersive spatial audio technique based on WFS has been discussed and validated with broadband virtual acoustic sources in normal reverberation conditions in a studio. Acoustic source localisation can effectively be achieved inside as well as outside the active control boundary by using the spatial audio technique. The ability allows users to use wider depth of virtual audio objects with other multimedia applications. The sound field generated by the virtual audio objects covers most of the domain over a sweet spot generated by conventional spatial audio applications.

In addition to the significant suppression of directional errors, the proposed method is also able to produce the desirable sound effectively even when there are some reverberations interfering with the sound and the system characteristics, e.g. impedance, are not known. The results prove the robustness and stability of the audio rendering system and demonstrate the potential advantages of the method in practical applications under these conditions required to implement a 3D multimedia system.

In many practical cases, e.g. spatial audio for home users, a bulky loudspeaker array surrounding all the walls is not suitable to be mounted permanently in a typical room. From a practical point of view, conventional WFS hardware, especially loudspeaker arrays, should be scaled down in size. In addition, a simple calibration method should be developed for user friendliness.

References

1. Theiler G, Wittek H (2004) Wave field synthesis: a promising spatial audio rendering concept. Acoust Sci Technol 25(6):393–399
2. Pierce AD (1991) Acoustics. An introduction to its physical principles and applications. McGraw-Hill Inc., New York, p 165

3. Merkle P, Karsten Müller, Aljoscha Smolic, Thomas Wiegand (2006) Efficient compression of multi-view video exploiting inter-view dependencies based on H.264/MPEG4-AVC. In: IEEE international conference on multimedia and expo 2006, Toronto
4. Remote Collaborative Real-Time Multimedia Experience over the Future Internet [online]. Available: http://www.ict-romeo.eu/
5. Nelson PA, Elliott SJ (1992) Active control of sound. Academic Press, San Diego, pp 118–22, 143–146, 311–378
6. Kincaid RK, Padula SL, Palumbo DL (1998) Optimal sensor/actuator locations for active structural acoustic control. AIAA Paper 98-1865, in Proceedings of the 39th AIAA/ASME/ASCE/AHS/ASC Structures, Dynamics and Materials Conference, Long Beach, CA
7. Spors S, Teusch H, Rabenstein R (2002) High-quality acoustic rendering with wave field synthesis. Vision, Modeling, and Visualization, pp 101–108
8. Williams EG (1999) Fourier acoustics, sound radiation and nearfield acoustical holography. Academic Press, San Diego
9. de Vries D (1996) Sound reinforcement by wave field synthesis: adaption of the synthesis operator to the loudspeaker directivity characteristics. J Audio Eng Soc 44(12):1120–1131
10. Sonke J-J (2000) Variable acoustics by wave field synthesis. Thela thesis, Amsterdam, Netherlands, ISBN 90-9014138-3
11. Ahrens J, Rabenstein R, Spors S (2008) The theory of wave field synthesis revisited. Audio Eng Soc Convention(124)
12. Lim H, Kim C, Hill AP, Ekmekcioglu E, Dogan S, Kondoz AM, Shi X (2014) An approach to immersive audio rendering with wave field synthesis for 3D multimedia content. In: The International Conference on Image Processing, Paris
13. Oldfield R, Drumm I, Hirst J (2010) The perception of focused sources in wave field synthesis as a function of listener angle. In: 128th AES Convention
14. Verheijen E (1998) Sound reproduction by wave field synthesis. Ph.D. thesis, Delft University of Technology
15. Spors S, Rabenstein R, Ahrens J (2008) The theory of wave field synthesis revisited. In: 124th AES Convention, Amsterdam, The Netherlands
16. Spors S, Kuntz A, Rabenstein R (2003) An approach to listening room compensation with wave field synthesis. In: Proc. AES 24th Int. Conf., Nuremberg, pp 70–82
17. RME HDSPe MADI [online]. Available http://www.rme-audio.de/en_products_hdspe_madi.php
18. Peters N, Lossius T, Schacher JC (2012) SpatDIF: principles, specification, and examples. In: Proc. SMC, vol. 20
19. Ahrens J, Spors S (2011) Wave Field synthesis of moving virtual sound sources with complex radiation properties. J Acoust Soc Am 130(5):2807–2816
20. 4189-A-021-1/2-inch Free-field Microphone [online]. Available http://www.bksv.com/Products/transducers/acoustic/microphones/microphone-preamplifier-combinations/4189-A-21
21. PULSE Analyzer Platform [online]. Available http://www.bksv.com/Products/pulse-analyzer
22. Moore BCJ (2003) An introduction to the psychology of hearing, 5th edn. Elsevier Academic Press, San Diego, pp 216–220
23. Wierstorf H, Raake A, Spors S (2012) Localization of a virtual point source within the listening area for wave field synthesis. In: 133rd AES Convention
24. Wittek H (2007) Perceptual differences between wave-field synthesis and stereophony. Ph.D. thesis, University of Surrey
25. Method for Subjective Assessment of Intermediate Quality Level of Coding Systems, Standard, ITU-R BS.1534-1 220-10 (2001–2003) The ITU Radio communication Assembly
26. Tamesuea T, Yamaguchib S, Saekib T (2006) Study on achieving speech privacy using masking noise. J Sound Vib 297:1088–1096

Chapter 12
Utilizing Social Interaction Information for Efficient 3D Immersive Overlay Communications

Theodore Zahariadis, Ioannis Koufoudakis, Helen C. Lelligou,
Lambros Sarakis, and Panagiotis Karkazis

Abstract Tele-immersive 3D communications pose significant challenges in networking research and request for efficient construction of overlay networks, to guarantee the efficient delivery. In the last couple of years various overlay construction methods have been subject to many research projects and studies. However, in most cases, the selection of the overlay nodes is mainly based on network and geographic only criteria. In this book chapter, we focus on the social interaction of the participants and the way that social interaction could be taken into account for the construction of a network multicasting overlay. More precisely, we mine information from the social network structures and correlate it with network characteristics to select the nodes that could potentially serve more than one users, and thus contribute toward the overall network overlay optimization.

12.1 Introduction

Advances in computer graphics, computer vision, 3D capturing and reconstruction and the combination of sensors and social networks show a path toward a 3D immersive internet, i.e. the internet that delivers tele-immersive experiences.

T. Zahariadis (✉)
Synelixis Solutions Ltd., Chalkida, Greece

Electrical Engineering Department, Technical Institute of Sterea Ellada, Psachna, Greece
e-mail: zahariad@synelixis.com

I. Koufoudakis
Synelixis Solutions Ltd., Chalkida, Greece

H.C. Lelligou • L. Sarakis
Electrical Engineering Department, Technical Institute of Sterea Ellada, Psachna, Greece

P. Karkazis
Electrical & Computer Engineering Department, Technical University of Crete,
Chania, Greece

© Springer Science+Business Media New York 2015   225
A. Kondoz, T. Dagiuklas (eds.), *Novel 3D Media Technologies*,
DOI 10.1007/978-1-4939-2026-6_12

Tele-immersion enables groups of geographically isolated users to interact and communicate in a realistic way, through real-time sharing of a virtual environment. Users and objects are captured by 3D cameras, which digitize them into a cloud of points or a 3D mesh. Moreover, virtual characters and objects can be generated and animated with computer graphics techniques. Digitized and compressed real and virtual users and objects are mixed into real reconstructed scenes or virtual worlds. Despite advances in 3D media compression and mechanisms such as caches that increase network efficiency [1], due to the large amount of exchanged data and the significant requirements for interactivity, 3D immersive communications pose severe challenges in networking infrastructure.

It is obvious that supporting real-time, network friendly 3D immersive communications requests for optimal, sometimes on demand, network overlay construction to guarantee the efficient content delivery. The last few years, overlays construction, especially in the form of Application Layer Multicasting (ALM) overlays to support stream multicasting, has been subject to many research projects and studies [2, 3]. However, these approaches are mainly based on network and geographic only criteria for the node selection and the overlay construction and do not take into account social interaction of the participants.

In this book chapter, we focus on the social interaction aspects and the way that social interaction of the participants could be taken into account for the construction of the overlay network topology. More precisely, we analyse information that we can mine from the social network structures and correlate it with the 3D immersive communications network overlay optimization process. We extract information from the interactivity between the participants, and their mobility behaviour to select the overlay nodes that could potentially serve more than one users and thus contribute toward the overall network optimization.

12.2   Network Overlay Categories

In typical immersive applications, the content exchanged may be grouped in two categories: (a) the static, which is pre-created and includes collections of digital objects (such as definitions of virtual worlds, background objects, 3D models, even stored Audio or Video streams) and (b) the dynamic, which consists of the content that is created and streamed "on-the fly", during the execution of the service, including control messages and 2D/3D video and audio in separate sessions. The percentage of the content between these two categories is not predefined and depends on the type of the application (e.g. interactive, social or serious games, or educational activities). A common approach to deliver these types of data is to create network overlays. Network overlays provide a higher level view on top of current network infrastructure, in a broader sense similar to Virtual Private Networks (VPN), ready to be deployed for specialized services. Overlay functionality typically focuses on QoS, fault tolerance, security, multicasting and content access, as depicted in Fig. 12.1.

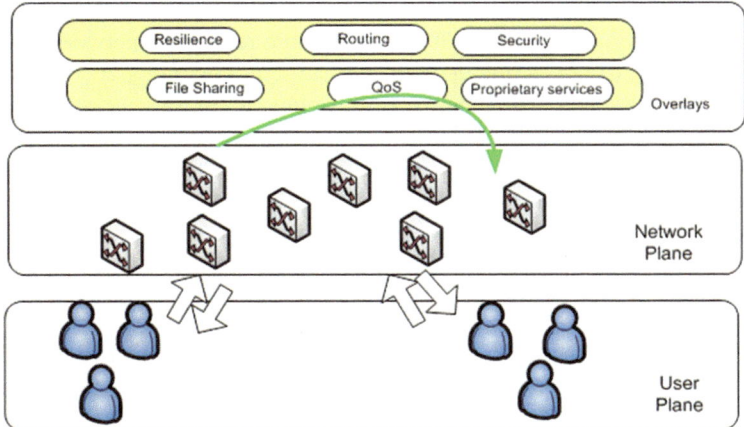

Fig. 12.1 Overlay networking

In order to cover both static and dynamic content, we consider two functionally separated overlays: (a) the *Distributed Media Overlay,* focusing on caching and delivering of static content and (b) the *Real Time Interactive Communication Overlay*, catering for the exchange of dynamic content and supporting signaling and monitoring for sessions, 3D meshes and point clouds, along with real-time video and audio.

Knowledge of the underlying network infrastructure and conditions (capacity, traffic, delay, load, network bottlenecks) as well as the network requirements of the 3D scene may significantly change the construction of the overlays and, consequently, the performance of the application and the Perceived Quality of Experience (PQoE). Moreover, the operational cost caused by the applications may be an issue for an operator, depending on the type of application. For instance, P2P file-sharing applications cause a considerable load on an operator's network and this happens over a long duration. Applications cannot usually judge the costs caused by an action, such as loading data from one or another location, as these costs are transparent to the applications.

In order to improve and even control the delivery of the content, the 3D immersive communications platform should be aware of the network capabilities, conditions and infrastructure in order to select the most appropriate participating nodes (i.e. routers, rendezvous points, ALM nodes, caches, application server(s), and clients) and build the most efficient overlay topology. This is a challenging objective from multiple points of view [4]. However, path selection based on background monitoring of alternative paths in an overlay of N nodes requires constant monitoring of $O(N^2)$ paths to choose the best detour at any given time [5]; as a result the efficiency and adoptability of this approach quickly diminishes with increasing number of nodes [6]. Moreover, having full knowledge of the network topology is out of scope of a tele-immersive platform. Yet, the smart selection of the overlay nodes may (a) increase the quality of the data path (e.g. reduce delay, jitter, etc.), (b) reduce the control overheads, (c) significantly increase the performance of the overall network and reduce the network bandwidth requirements and (d) reduce the network infrastructure operational costs.

In the literature, there are many overlay construction algorithms that try to optimize various criteria such as to minimize the average cost of shortest path trees rooted at group members, minimize the overlay topology diameter while respecting the degree constraints, minimize the root-path latency for a specific node, minimize the average delay and maximize the average bandwidth. We summarize and propose the following high level algorithm:

- **Step 1:** Network nodes, which are members of the overlay will keep a connected mesh topology among themselves following one of the many approaches available in the literature [3]. The mesh topology is explicitly created at the beginning of a session, based on various "cost" metrics.
- **Step 2:** Whenever a new client joins a session, a spanning tree will be created over the mesh topology and will be optimized based on a number of criteria (mainly avoidance of loops, while keeping the minimal latency and the average bandwidth requirements).
- **Step 3:** Depending on the assumed life-time of the sessions, a *Spanning Tree refinement* procedure will periodically run to improve the overlay performance as well as to deal with fluctuations in the available network resources and congestion. Yet, as the effectiveness of the refinement to real-time applications is questionable, due to interrupted data distributions among the members, this approach will be further studied at deployment time.

In all three steps, the "cost" of an overlay path between any pair of nodes should be comparable to the "cost" of the unicast path between that pair of members.

12.3 Constructing the Network Overlay

In order to construct an overlay topology various direct and implicit (network related, node related, network operator related, social network related) metrics should be considered to select the "best" nodes for each client per session or the "best" parent nodes, during the construction of a multicast overlay. As "best" node, we define the node that provides the optimal average PQoE to the 3D immersive communications clients that participate in the session, with minimal network overhead and operational cost. In parallel, the proposed solution should avoid approaches that drastically increase the network load (e.g. such as P2P techniques) and decrease the overall efficiency of the network. Last but not least, we should take into account, metrics related to the cost of the network, if applicable. As a result, a number of parameters should be taken into account, such as:

(a) *Network Characteristics.* We assume a number of metrics such as the end-to-end available bandwidth, the end-to-end network delay, one-way delay, potential network bottlenecks and the delay jitter. *Active probing* or passive monitoring techniques may be applied to mine the network characteristics. These metrics may have quite different result, based on the type and characteristics of content that needs to be delivered [10].

(b) **In-network Nodes Characteristics.** The nodes (both network nodes and client nodes participating in the immersive communication) are not equivalent, either due to their different characteristics (e.g. CPU capacity, storing capacity/memory, buffering queues, network interfaces) or due to their role in the network. As a result, the combination of network's characteristics and nodes' characteristics defines the technical metrics, which could be measured. As the client nodes are given, we may only select alternative in-network nodes that have less load or higher capabilities. From these two categories, we can also extract information about potential network bottlenecks.

(c) **Network Operator recommendations.** Besides performance and technical characteristics, there may be a set of other parameters and recommendations, provided directly or indirectly by the network operator/ISP that needs to be taken into account. For example, by utilizing information from an IETF ALTO server [7], information such as the ISP communication cost, inter-ISP contracts/agreements and network related policies may be taken into account.

(d) **Social Interaction.** As already stated, besides pure network and nodes characteristics, the network overlays may be set-up taking into consideration the social dimension of the social interaction of the 3D immersive communication users (e.g. [9]); i.e. based on the frequency and duration that specific users participate in a common tele-immersion session, we may modify clients' point of connection.

In the next section, we focus on the social interaction. It should be emphasized that the purpose is not to find the best mathematical equation in order to model the various measurements and probability, but to compromise between valid results and fast calculations, with low computational complexity.

12.4 Social Network Analytics for Overlay Construction Optimization

In many network overlay optimization methodologies and algorithms, the overlay is considered as a graph, where nodes are the elements positioned according to their location, while links represent communication paths through intra-net or internet layouts, being quantified by weight values (e.g. capability primitives and communication cost). One of the key issues of social networking is that social interaction can be actually recorded and processed using big data analysis frameworks in order to remap the user population on a social network canvas. Hence, analytics from the social network can produce behavioural semantic relational graphs and clustering at the user level. By exploring the social interactions of users, we could unveil their social relationships and identify potential communities, while building a social profile of both individuals and communities. This will create overlay social network links and relations, based on pure social semantic primitives. This social network can be further analysed [11, 12, 16] in order to:

- Detect the social communities with interacting session patterns.
- Build a social profile for the individual user: social connectivity, interaction topics, number of communities the user belongs to, etc.
- Build a social profile for communities: number of members, group cohesion, etc.
- Profile the communities across a number of attributes (content access, session primitives, socio-demographics...) based on the interrelation principles that appear in social groups.

By projecting social communities to the network paradigm and narrowing the social interaction to a more system-scoped level (enabled by technologies like [17] and [18]), community formulation could be focusing on users accessing the overlay through common session topics with the following relation criteria:

- Sharing audio streams
- Sharing 2D or 3D video streams
- Sharing live streams through interaction and collocation in virtual worlds

In order to better explain the issue, let us take as an example the network topology shown in Fig. 12.2. Blue nodes are end users and grey nodes are network nodes.

Let us assume that end-user A aims to communicate with end-user D. Assuming that all network nodes and links are equivalent, from a networking point of view, the best path is:

$$A \leftrightarrow 1 \leftrightarrow 2 \leftrightarrow 6 \leftrightarrow 9 \leftrightarrow D$$

Now, let's assume that B and C are very close friends of A and/or D, and based on the recent history of the social interaction, there is a high probability that B and/or C will join the session in few minutes. In that case, maybe instead of the previous path, it is better to select the path:

$$A \leftrightarrow 1 \leftrightarrow 5 \leftrightarrow 7 \leftrightarrow 8 \leftrightarrow 9 \leftrightarrow D$$

Though this path is suboptimal, and it is longer by one link as compared to the previous one, it is at the end a better path, if B and/or C join the session, as content that already streams via, or is cached at nodes 5 and 7, may be directly multicast or be forwarded to clients B and/or C via nodes 4 and 10 respectively. However, we also need to take into account mobility, as if clients B and/or C have a high mobility profile, they may join from other nodes, reducing the overall benefit of the proposed path selection.

We argue that the above challenge is facilitated by user community/cluster detection through social network analysis. Assuming limited nomadic user patterns, these node clusters can then be mapped to the network topology. The next step is to

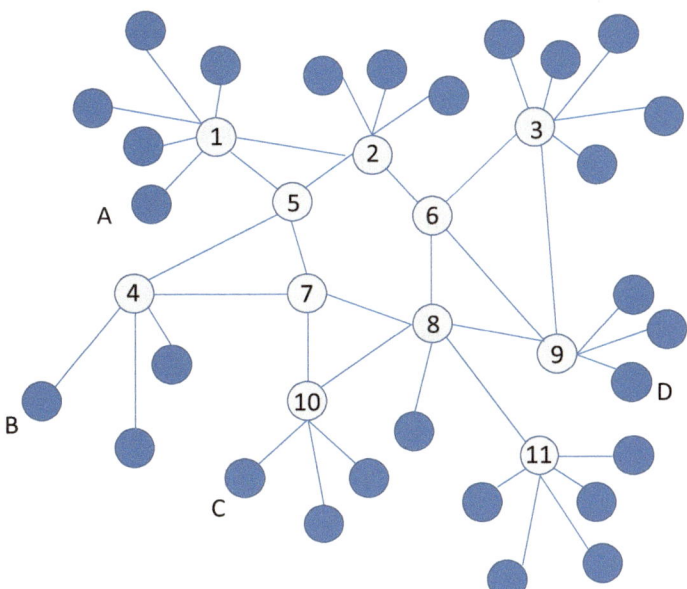

Fig. 12.2 Network topology example. End users A–D form a social network

apply classical routing algorithms, such as spanning tree creation/refinement, in order to identify the optimum routes of service delivery (ALM service nodes, cache content placement nodes) that would best service the identified user community. The operation rationale behind this concept at service level is to have background processes implementing the above workflow, identifying social communities and pre-calculating for them routing trees, cache them and deploy them when temporal or session-driven trigger conditions are met (e.g. login of $N1$ out of N members of the community).

The main advantage in this approach is that it breaks down the problem of identifying optimum content placement strategies and multicast trees, to multiple smaller problems involving, small groups of nodes, of which the statistical probability of its usefulness can be quantified based on historic data. Furthermore, data processing and algorithm execution is asynchronous without real-time requirement. It is a task executed in the background and could be delegated to distributed computing frameworks, such as Map Reduce [13–15] containers. It is worth mentioning that open platforms such as Apache Hadoop (MapReduce container [13]) and Apache Mahout (Data Mining Big Data Engine [14]) have evolved significantly so that this level of processing complexity can be feasible at a high programming level and low investment cost.

12.4.1 User Interaction Probability

Let us assume that in a social network, there are N_t types of interaction between N_u users (e.g. chat, talk, like). A simple interaction probability Ip^{u_1,u_2} from user u_1 to user u_2 for a time period t could be defined as:

$$Ip^{u_1,u_2} = \sum_{m=1}^{Nt} w_m * \frac{N_m^{u_1,u_2}}{N_m^{u_1}} \tag{12.1}$$

where $N_m^{u_1,u_2}$ is the number of interactions of type m from user u_1 to user u_2, $N_m^{u_1}$ is the number of interactions of type m from user u_1 to any other user and w_m is a weight factor showing the importance of interaction of type m, where:

$$\sum_{m=1}^{Nt} w_m = 1 \tag{12.2}$$

As many sessions could have more than two participants, it is:

$$N_m^{u_1} \neq \sum_{k=1}^{Nu} N_m^{u_1,u_k} \tag{12.3}$$

This probability metric can determine the level of "connectedness" between users u_1 and u_2. It is worth noting that in general, $Ip^{u_1,u_2} \neq Ip^{u_2,u_1}$, as relationships may not always be symmetrical, e.g. if u_1 has established an immersive session, it is very possible that also u_2 will be connected. However, u_2 being more social may have sessions with other persons, without u_1 participation. Moreover it is important to note that interactions between more than two users have not been taken into account. This is also an important factor, but it goes well beyond the scope of this chapter.

12.4.2 User Mobility Probability

In order to put this approach into effective action, we need also to take into account the mobility patterns of users. Users that connect to the 3D immersive network from continuously changing network locations are not suitable candidates for this kind of cluster-based optimization. This can be represented by a Connectivity Variation probability a.k.a. how often a user is connected from the same location.

Let us assume that in a time period t, a user u_1 is connected N_{u1} times from N_n different geographical nodes/locations. The probability Cp^{u_1,n_1} of u_1 to be connected via node/location n_1 is

$$Cp^{u_1,n_1} = \frac{N_{u_1,n_1}}{N_{u_1}} \tag{12.4}$$

where N_{u_1,n_1} is the number of times that u_1 has been connected to the 3D immersive communications framework via node n_1. Aging factors and Markov chains could also be introduced, but would further complicate the approach. A simplified Location Connectivity Variance $LCVp(u_1)$ of user u_1 could be defined as:

$$LCVp^{u_1} = 1 - \sqrt{N_n \times \frac{\sum_{i=1}^{Nn}(Cp^{u_1,i} - Cp^{u_1})^2}{N_n - 1}} \tag{12.5}$$

where N_n is the number of different geographical nodes/locations that user u_1 has been connected to the social network and Cp^{u_1} is the average probability of u_1 to be connected via any node/location. Equation (12.5) can also be written as:

$$LCVp^{u_1} = 1 - \sqrt{N_n \times s_{u_1}} \tag{12.6}$$

Where s_{u_1} is the corrected sample standard deviation of the average probability of u_1 to be connected via any node/location. As a result, $1 - LCVp^{u_1} = \sqrt{N_n \times s_{u_1}}$ is the probability that u_1 remains at the same network location (the same geographical location/node).

Based on the above, we define as **Location Dependent Interaction probability** $LDIp^{u_1,u_2}$ the probability of u_1 to interact with u_2, having both u_1 and u_2 static as:

$$LDIp^{u_1,u_2} = Ip^{u_1,u_2} \times (1 - LCVp^{u_1})^a \times (1 - LCVp^{u_2})^a \tag{12.7}$$

where α is a small integer >1 that shows the importance that we would like to give to the location connectivity pattern of the users. The $LDIp^{u_1,u_2}$ can be used for the overlay hierarchy optimization, as it directly associates the users u_1 and u_2.

12.5 Node Selection Based on Social Interaction

Taking into account the social network interaction analysis and connectivity variation pattern of users, we may select a node A as having an optimal location to serve as overlay node with respect to node B based on the following algorithm:

- **Step 1**: Based on Eq. (12.4), select a group U_1 with the k_1 users that most often connect to the overlay using node B as access node.

$$U_1 = \left\{ u_a \left| Cp^{u_a,B} \in \max_{j=1\ldots k_1} \left(Cp^{u_j,B} \right) \right. \right\}$$ (12.8)

- **Step 2**: Based on Eq. (12.7), for each user u_a of U_1, select a group U_2 with the k_2 users that have the strongest and most often interaction with user u_a and their connectivity variation pattern shows that they are most often connected from a given location.

$$U_2 = \left\{ u_b \left| LDIp^{u_a,u_b} \in \max_{u_a \in U_1, j=1\ldots k_2} \left(LDIp^{u_a,u_j} \right) \right. \right\}$$

- **Step 3**: Based on Eq. (12.4), for each user u_b of U_2, select a set X_n with the n nodes that most often user u_b connects to the framework using this node as access node.

$$X_n = \left\{ x_i \left| Cp^{u_b,x_i} \in \max_{u_b \in U_2, j=1\ldots n} \left(Cp^{u_b,x_j} \right) \right. \right\}$$ (12.9)

- **Step 4**: If we define $NSp^{A,B}$ the *Node Selection probability* for selecting the node A from a node B as an optimal location for an overlay node taking into account only the network criteria, we may select node A from node B based on the social interaction of the users in case:

$$\varphi_1 NSp^{A,B} + \varphi_2 NSp^{A,x_i} > SNS$$ (12.10)

where SNS is a threshold, ϕ_1 and ϕ_2 are weighting factors to give emphasis to the social dimension of the node B and x_i, and

$$\phi_1 + \phi_2 = 1$$ (12.11)

Equation (12.10) may be also used for the overall network overlay optimization.

12.6   Validation

In this section, we provide an initial validation of the method that has been presented. For the evaluation of the interaction probability, we use three services: chat, talk and video conference with weights w_c, w_t and w_v respectively and realize the following scenarios (see Table 12.1):

In all scenarios, we give more value to the video conference metric, as people that are used to communicate via video conference would be at least the early adopters of the 3D immersion communications system. Users that mainly chat and not use voice or video conference today may not utilize it in the near future.

Table 12.1 Social network service weights per scenario

	Scenario 1	Scenario 2	Scenario 3	Scenario 4	Scenario 5
Chat	0.3	0.2	0.3	0.2	0.1
Talk	0.3	0.3	0.2	0.2	0.1
Video conf.	0.4	0.5	0.5	0.6	0.8

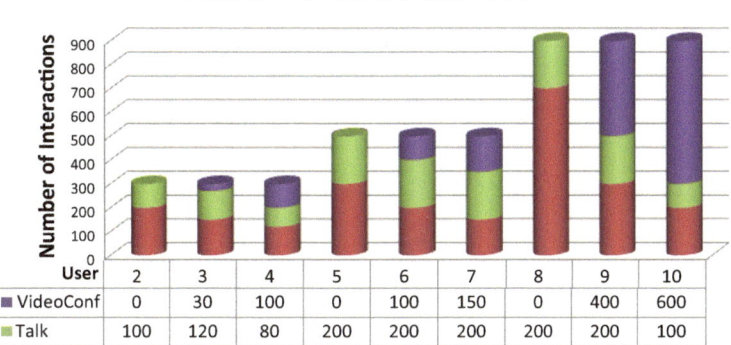

Fig. 12.3 Number of Interactions between User 1 and Users 2–10

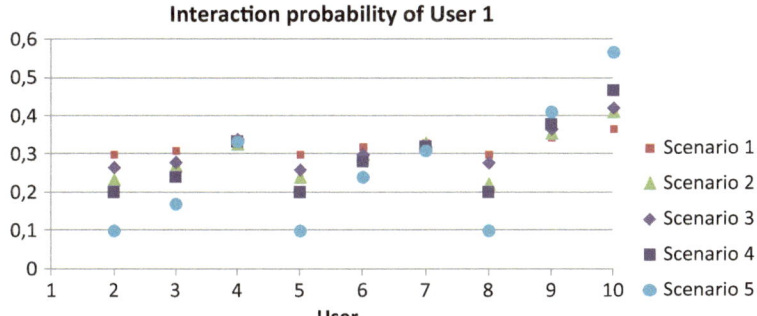

Fig. 12.4 Interaction probability of User 1

In Fig. 12.3, we summarize some indicative interactions between User 1 and Users 2–10. We have selected the values in order to reflect different communication volumes and different media types.

In Fig. 12.4, we show the interaction probability of User 1 with Users 2–10 calculated using the Eq. (12.1). It can be seen that user 10, who has the largest number of video conference events with user 1, has also the highest probability to interact.

In Fig. 12.5, we show the location connectivity behaviour of Users 1–10. We assume that there are 4 locations/nodes that each user could be connected at and we

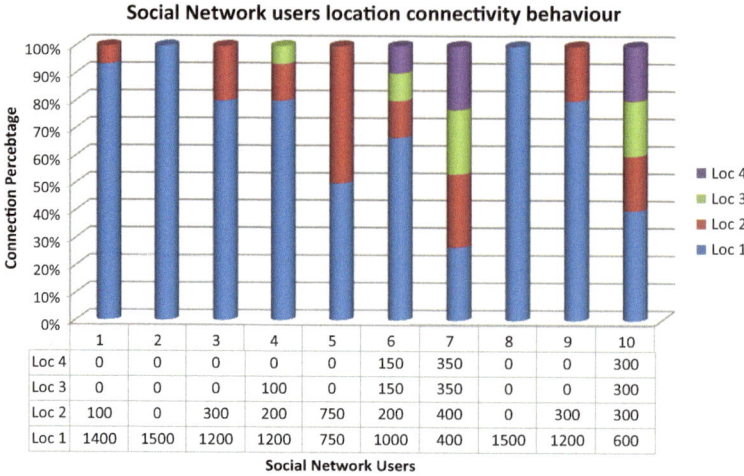

Fig. 12.5 Number of connections (login) via various nodes/locations

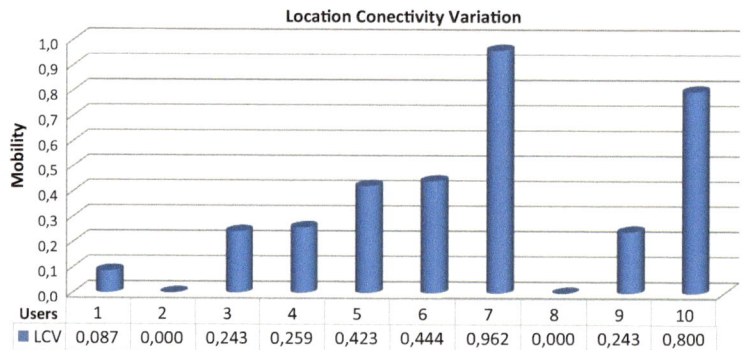

Fig. 12.6 Users' location connectivity

represent various mobility patterns. Of course the connectivity location 1 of User 1 may be different from the connectivity location 1 of User 2. Besides User 1, who is our reference user, we have split the remaining users into two groups. In the first group (Users 2–7) that shows different patterns of location connectivity behaviour and the second group (Users 8–10) that shows three users with radically different location connectivity behaviour to be used for the overall comparison of the Location Dependent Interaction (LDI) graphs.

In Fig. 12.6, we show the users' location connectivity variation calculated based on Eqs. (12.5) and (12.6). We notice that user 2 and user 8 have zero mobility, while user 7 that may connect from any of the 4 locations with equal probability has a mobility that is approaching 1.

User 3 has slightly smaller mobility than user 4, as s/he connects with the same probability from location 1, but besides that, user 3 connects only from location

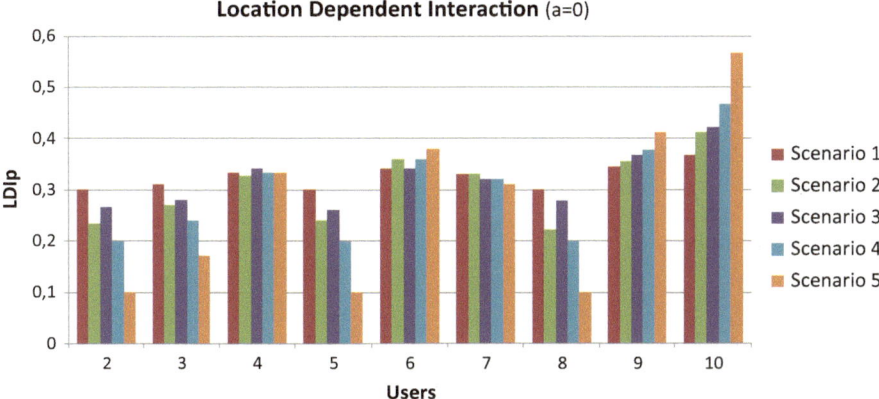

Fig. 12.7 Location dependent interaction ($\alpha = 0$)

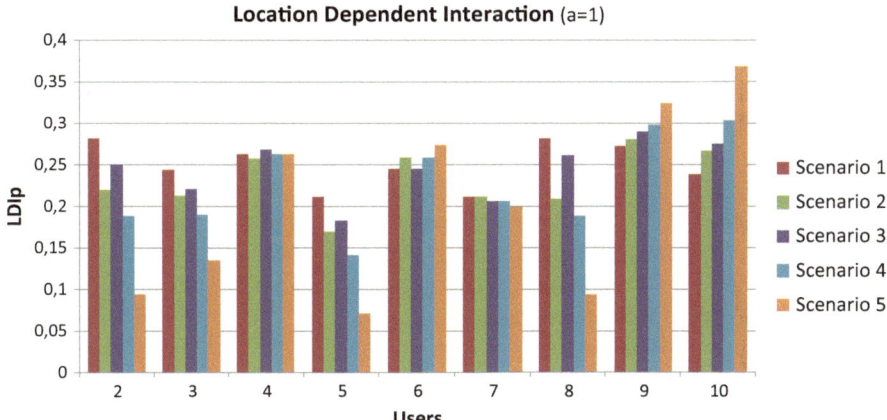

Fig. 12.8 Location dependent interaction ($\alpha = 1$)

2, while user 4 connects from both locations 2 and 3. Similarly, the difference in the location connectivity variation between user 5 and 6 is small, as user 5 connects only from two locations but with equal probability, while user 6 connects from 4 different locations, but still location 1 is by far the main connectivity location. User 8 is a user that has zero mobility, user 9 is a user that mainly connects from one location and user 10 is a user that may connect from many locations with high probability.

By applying the above data in Eq. (12.6), we get the location dependent interaction graphs (Figs. 12.7, 12.8, and 12.9).

In the Fig. 12.7, we have applied $\alpha = 0$, thus the location connectivity variation behaviour is filtered out. We can see that location dependent interaction probabilities for User 9 and 10 are clearly higher, due to their interaction probabilities,

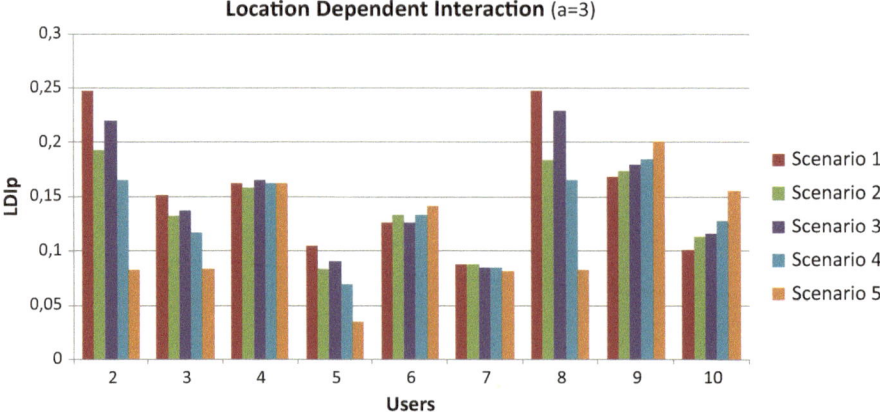

Fig. 12.9 Location dependent interaction ($\alpha = 3$)

especially in scenarios 4 and 5 that put more emphasis on the video conference interactions.

In Fig. 12.8, we have applied $\alpha = 1$, thus the location connectivity variation behaviour is taken into account but with limited importance. Location dependent interaction probabilities for user 9 and user 10 remain high, especially in scenarios 4 and 5, but also user 5 is considered, as the interaction model is not very good, but s/he is very stable with respect to the connectivity probability.

Finally, in Fig. 12.9, we have applied $\alpha = 3$, thus the location connectivity variation behaviour is seriously taken into account. Location dependent interaction probabilities for user 9 and user 10 are now much lower, while user 5 has an increased location dependent interaction probability as s/he is very stable as far as the connectivity probability.

In general, we believe that $\alpha = 3$ is closer to the value that needs to be selected, as we consider the location connectivity variation quite important for the network overlay nodes selection.

Conclusions

Tele-immersive 3D communications pose significant importance in the network delivery. We have investigated the benefits from network awareness for the provision of networked content delivery and optimization of the network overlays. While there are multiple benefits, in the context of Tele-immersive 3D communications, we may highlight the following:

(a) The awareness of network topology and of its characteristics can guide content providers to tailor the quality of the content offered to the end users.

(b) The infrastructure provider can better utilize auxiliary resources, including the activation and operation of replication servers (caches).

(continued)

(continued)

Moreover, 3D immersive communications pose significant importance in the multicasting delivery. As IP multicasting, peer-to-peer networking and content-centric networking have not been sufficiently deployed (and there are even doubts if they will ever be), Application Layer Multicasting (ALM) may provide the required functionality. However, the known in the literature approaches for ALM nodes selection and ALM overlay (tree or mesh structure) construction are mainly based on network and geographic only criteria.

In this book chapter, we have shown that by monitoring and analysing the Social Network interaction of the participants, we can extract a new perspective for network awareness that can be exploited in order to correlate network planning and algorithmic optimizations with digests and primitives of social network analysis. In this way, user behaviour may be mapped in the applications' framework and social interaction can be reflected in multicasting and routing algorithms and content delivery topologies.

Acknowledgments This work has been partially co-financed by the European Commission through the FP7-ICT-287723 REVERIE project.

References

1. Papadakis A, Zahariadis T, Mamais G (2013) Advanced content caching schemes and algorithms. Poznan Univ Technol: Adv Electronics Telecommun 3(5):10–16
2. Yeo CK, Lee BS, Er MH (2004) A survey of application level multicast techniques. Elsevier, Comput Commun 27:1547–1568
3. Hosseini M, Ahmed D, Shirmohammadi S, Georganas N (2007) A survey of application-layer multicast protocols. IEEE Commun Surv 3rd Quarter 9(3):58–74
4. Papadakis A, Zahariadis Th, Vrioni C (2014) Monitoring and tailoring the usage of network re-sources for smoothing the provision of interactive 3D services. In: IEEE International Conference on Image Processing (ICIP 2014), Paris, 27–30 Oct
5. Subramanian L, Stoica I, Balakrishnan H, Katz R (2004) OverQoS: An Overlay Based Architecture for Enhancing Internet QoS, Usenix NSDI, March
6. Wang G, Zhang B, Eugene Ng TS (2007) Towards network triangle inequality violation aware distributed systems. ACM IMC, San Diego
7. http://datatracker.ietf.org/wg/alto/charter
8. IETF RFC 6455 (2011) The WebSocket Protocol
9. Zahariadis T, Papadakis A, Nikolaou N (2012) Simulation framework for adaptive overlay content caching. In: 35th IEEE International Conference on Telecommunications and Signal Processing (TSP), Prague, Czech Republic, 3–4 July
10. Vern Paxson (2004) Strategies for sound internet measurement. In: Proceeding IMC '04 Proceedings of the 4th ACM SIGCOMM conference on Internet measurement
11. Tom Heath (Talis), Christian Bizer (Freie Universität Berlin), Linked Data: Evolving the Web into a Global Data Space. Available at http://linkeddatabook.com/editions/1.0/

12. Christian Bizer, Tom Heath et al (2008) Linked Data on the Web. In: WWW Conference, Beijing, April
13. Apache Hadoop Wiki. http://wiki.apache.org/hadoop/
14. Apache Mahout Wiki. https://cwiki.apache.org/MAHOUT/mahout-wiki.html
15. Donald Miner, Adam Shook (2012) MapReduce design patterns, building effective algorithms and analytics for Hadoop and other systems. O'Reilly Publishing
16. https://drupal.org/
17. FP7/ICT FI-WARE, Future Internet Core Platform, "FI-WARE Data/Context Management". http://forge.fi-ware.eu/plugins/mediawiki/wiki/fiware/index.php/FI-WARE_Data/Context_Management
18. http://www.hpl.hp.com/research/linux/httperf/

Index

© Springer Science+Business Media New York 2015
A. Kondoz, T. Dagiuklas (eds.), *Novel 3D Media Technologies*,
DOI 10.1007/978-1-4939-2026-6